Silke Goronzy

Robust Adaptation to Non-Native Accents in Automatic Speech Recognition

 Springer

Series Editors

Jaime G. Carbonell, Carnegie Mellon University, Pittsburgh, PA, USA
Jörg Siekmann, University of Saarland, Saarbrücken, Germany

Author

Silke Goronzy
Sony International (Europe) GmbH, SCLE, MMI Lab
Heinrich-Hertz-Straße 1, 70327 Stuttgart, Germany
E-mail: goronzy@sony.de

Cataloging-in-Publication Data applied for

A catalog record for this book is available from the Library of Congress.

Bibliographic information published by Die Deutsche Bibliothek.
Die Deutsche Bibliothek lists this publication in the Deutsche Nationalbibliografie;
detailed bibliographic data is available in the Internet at <http://dnb.ddb.de>.

CR Subject Classification (1998): I.2.7, I.2, J.5, H.5.2, F.4.2

ISSN 0302-9743
ISBN 3-540-00325-8 Springer-Verlag Berlin Heidelberg New York

Springer-Verlag Berlin Heidelberg New York,
a member of BertelsmannSpringer Science+Business Media GmbH

http://www.springer.de

© Springer-Verlag Berlin Heidelberg 2002
Printed in Germany

Typesetting: Camera-ready by author, data conversion by Boller Mediendesign
Printed on acid-free paper   SPIN: 10871801   06/3142   5 4 3 2 1 0

Lecture Notes in Artificial Intelligence 2560

Subseries of Lecture Notes in Computer Science

Edited by J. G. Carbonell and J. Siekmann

Lecture Notes in Computer Science

Edited by G. Goos, J. Hartmanis, and J. van Leeuwen

Springer
Berlin
Heidelberg
New York
Hong Kong
London
Milan
Paris
Tokyo

Foreword

The present state of the art in automatic speech recognition - under certain conditions - permits man-machine-communication by means of spoken language. Provided that speech recognition is tuned to the common native language (target language) of the users, speaker-independent recognition of words from a small vocabulary is feasible, especially if the words are spoken in isolation, and for larger vocabularies at least speaker-dependent recognition performance is satisfactory. The most elaborate up-to-date speech recognition systems manage large vocabularies even in speaker-independent connected speech recognition. However, perfect zero error rates cannot be achieved, and therefore such systems can be used only in applications that to some degree are fault tolerant, e.g. within dialog systems that offer appropriate feedback to the user and allow him to correct recognition errors in a convenient way.

Generally, error rates for speaker-dependent recognition are lower than for speaker-independent recognition, and error rates between these limits are achieved by systems that by default are preset to speaker-independent recognition and can be tailored to a certain speaker by means of a more or less demanding adaptation procedure. For adaptation purposes the prospective user of the system is required to produce a rather large number of certain prescribed utterances. Although some speaker-adaptive systems of this kind are already available, it is still a matter of research, as to how the required number of utterances can be reduced, in order to make the adaptation procedure less demanding for the prospective user. However, at several places more ambitious research work is directed towards speech recognition systems that continuously adapt to the current speaker without requiring a separate adaptation phase at each change between speakers. The work that is presented here is a major step towards such a system and - due to a remarkable new approach - adaptation can be successful even in the case of a non-native speaker with a foreign accent.

In an elaborate speech recognition system there are several knowledge sources among which the pronunciation dictionary is of special interest for measures against non-native accents. Usually, this dictionary denotes the native pronunciations for all items of the vocabulary. In oder to make the system ready for non-native accents, also the respective non-native pronunciations must be entered into the pronunciation dictionary. Up to now these additional entries either had to be specified by an expert for the special pair of target language and foreign language or had to be extracted automatically from a large number of spoken examples. The new approach is based upon the idea of processing the vocabulary and pronunciation dictionary of the target language with special attention to the phoneme inventory of the foreign language. Thus, any desired pair of target language and foreign language can be conveniently managed without the help of special experts and without any spoken examples. This is the most important among the contributions of this work to the fundamentals for the development of future speech recognition systems.

November 2002   E. Paulus

Preface

Speech recognition technology is being increasingly employed in human-machine interfaces. Two of the key problems affecting such technology, however, are its robustness across different speakers and robustness to non-native accents, both of which still create considerable difficulties for current systems.

In this book methods to overcome these problems are described. A speaker adaptation algorithm that is based on Maximum Likelihood Linear Regression (MLLR) and that is capable of adapting the acoustic models to the current speaker with just a few words of speaker specific data is developed and combined with confidence measures that focus on phone durations as well as on acoustic features to yield a semi-supervised adaptation approach. Furthermore, a specific pronunciation modelling technique that allows the automatic derivation of non-native pronunciations without using non-native data is described and combined with the confidence measures and speaker adaptation techniques to produce a robust adaptation to non-native accents in an automatic speech recognition system.

The aim of this book is to present the state of the art in speaker adaptation, confidence measures and pronunciation modelling, as well as to show how these techniques have been improved and integrated to yield a system that is robust to varying speakers and non-native accents.

The faculty of electrical engineering of the Technical University Carolo-Wilhelmina of Braunschweig has accepted this book as a dissertation and at this point I would like to take the opportunity to thank all those who supported me during this research work. First of all I would like to thank Prof. Dr.-Ing. Erwin Paulus for his valuable suggestions and his support. Likewise I would like to thank Prof. Dr.-Ing. Günther Ruske from the Technical University of Munich for giving the second opinion.

The research presented in this book was conducted while I was working at Sony in Stuttgart and I would like to express my gratitude that I got the permission to conduct and publish this research work.

I am also indebted to my colleagues, who supported me a lot by being very cooperative and helpful. I would like especially to thank Dr. Krzysztof Marasek and Andreas Haag for their co-operation in the field of confidence measures, Dr. Stefan Rapp for the valuable discussions and Dipl. Ling. Manya

Sahakyan for conducting the accompanying experiments in the area of pronunciation modelling.

Furthermore I would like to thank Dr. Elmar Noeth, Dr. Richard Stirling-Gallacher, Dr. Franck Giron and particularly Dr. Roland Kuhn for their many very helpful comments, which played an important part in improving this book.

Special thanks go to Dr. Ralf Kompe, whose support was an important contribution to this book and who, in spite of his many other duties, always found time to support me.

Most especially, I would like to thank my dear parents, Waldemar and Renate Goronzy, without whose affectionate upbringing my entire education and this book would not have been possible and who provided me with a set of values I could not have gained through any other academic education process.

Fellbach, September 2001   Silke Goronzy

Table of Contents

1 Introduction

In a recent interview Ray Kurzweil stated that in the year 2019 a $1000-computer will have computational power that is comparable to the human brain [Spi00]. This means it would be able to perform around 20 million billion operations per second. In the year 2029 he supposes software for 'intelligence' will be available that offers the capability to organise its own knowledge and share it with other computers more easily than humans can. One PC will then be doing the work of 1000 human brains. The important question however remains unanswered, what makes the consciousness of a human being? Each human is an individual, having his own understanding of certain things. If intelligence is considered to be the capability to solve (recurring) problems in a limited amount of time, then computers are already better than humans.

It is far beyond the scope of this book to answer the question whether computers will be able to achieve human performance and behaviour in the future, but it focuses on the methods and problems that occur, if computers are used in a human-machine conversation, which is only one of the tasks that are fulfilled so easily by humans in their every day life.

The conversation between human beings is a very complex situation. People can communicate intentions very easily without much effort, provided they are speaking the same language. Communication and language is very natural, the capability of which is acquired step by step, beginning when one is a baby and then growing and developing unconsciously throughout life.

When humans need to interact with computers however, the two partners often do not speak the same language. So far, computers often understood only certain commands and only if they were typed on a keyboard or a mouse was used. This special way of communication is in contrast to the human way of communication and it is one of the reasons why many people are afraid of using computers. But once humans understand how computers work, it is easy for them to learn the 'language' of the computer.

What we are interested in, however, is, how can we make computers understand the human language?

If we were able to solve this problem, the user could benefit from that. Today's information technology is rapidly changing, often too rapidly for the majority of the users to keep track of. An often mentioned example is the video cassette recorder (VCR) that many people have difficulties to program correctly. The main reason is that the user interfaces (UIs) are not intuitive and often only understandable to people with technical skills. But, using speech, the required function can very easily be formulated in one simple sentence, which could e.g. be 'I would like to record the thriller tonight on channel one'. The way that this information currently has to be entered to the VCR to make it record the right movie is far more complicated. This is a very simple example, but there is a variety of others. More and more people argue that the UIs for various kinds of devices need to be more intuitive and should lean towards human-human communication.

One step towards this is a technology that is called automatic speech recognition (ASR). It is an attempt to recognise human speech with machines. Applications already exist that employ this technique for controlling devices by accepting spoken commands. Other examples are interactive voice response (IVR) systems that are often used in telephone call centres to direct users to the right person or to let the user fulfil easy transactions on their bank account. Also information retrieval systems exist, e.g. train time-table query systems or cinema information systems; PC dictation systems that allow dictating letters or documents, are also already on the market.

Another reason that suggests the use of speech interfaces, is related to safety. All mobile applications, e.g. mobile phones and car navigation systems, impose a high safety risk when the driver of a car operates such a device while driving. Using the conventional menu-driven UI, the driver has to concentrate on the screen device and cannot simultaneously concentrate on driving. Speech interfaces would enable him to keep his hands on the wheel and his eyes on the street.

Of course the overall goal is to let computers understand human speech as it is used in human conversations. For solving this problem there are many technical difficulties that may seem to be no problem at all in human-human communication.

To give an impression of one of the occurring difficulties, consider the following example: When trying to understand a spoken sentence, humans do not care about word boundaries. In fact, in spontaneous speech there are often no pauses between the words. Figures 1.1 to 1.3 illustrate this difficulty.

It is hard to recognise speech

Fig. 1.1. Sentence typed with spaces between the words

If written normally, with spaces between the words, as shown in Figure 1.1, people have no difficulties in reading the sentence. This would correspond to a sentence spoken with long pauses between the words. In this case the ASR system could easily detect the pauses and thus the starting and ending points of the words.

Fig. 1.2. Sentence typed without spaces between the words

However, speaking a sentence with long pauses in between the words is very unnatural and usually people do not speak that way. The same sentence is shown in Figure 1.2, this time written without spaces between the words. It can be seen that it is much harder to read this sentence than the previous one.

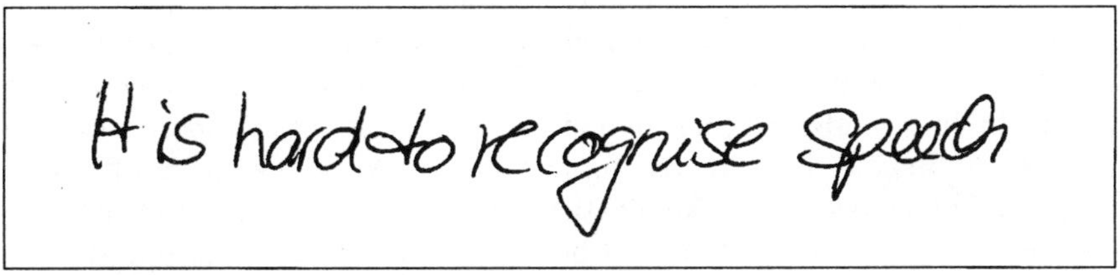

Fig. 1.3. Sentence in hand-writing

We know that each human has his own voice and often we can recognise a person just by hearing his voice. The sound of the voice does not influence our capability to understand what a person says, even though different voices of different people might have very different characteristics. To return to our example, the personal voice is analogous to the hand-writing of a person, which is also unique to each person. If we consider the hand-written counterpart of our sentence as shown Figure 1.3, we can easily see that it is even harder to read.

This simple example shows only vaguely the problems researchers are faced with when trying to automatically recognise speech. There are of course many more.

1.1 Outline of This Book

Among the many problems that are still unsolved, the research presented here is concerned with robust adaptation to native as well as to non-native speakers.

For many applications the problem of insufficient amounts of data available for the adaptation process arises. This topic is investigated in detail here and the known speaker adaptation method of Maximum Likelihood Linear Regression (MLLR) is improved to deal especially with very limited amounts of adaptation data in online adaptation, yielding the new 'weighted MLLR' approach.

Unsupervised adaptation is often preferable to supervised adaptation, because the user is not forced to undergo an enrolment phase. However, in unsupervised adaptation we are faced with misrecognitions that might impose problems on the adaptation. As a solution to this problem, a confidence measure (CM) is introduced that tries to judge how reliably an utterance was recognised by the ASR system. This can be compared to humans who often realise that they probably have not understood something correctly. The CM guides adaptation, yielding a semi-supervised adaptation, so that only reliably recognised utterances are used. For the computation of this CM we introduce new features that are capable of improving the classification rates compared to traditionally used features.

While most systems focus on native speakers of a certain language, non-native speakers are a big problem. Due to growing globalisation, the increased mobility and the growing Internet, all of which enable people to 'go around the world' in seconds, the topic of non-native speech becomes more and more important. Since it is currently impractical to design speech interfaces for all possible languages, each application has to be restricted to use the most common ones. So there will always be a vast number of people who will use one of these languages without this language being their mother tongue. That means they may speak, e.g., English with a non-native accent. Especially public information systems and Internet applications will have to deal with this problem.

However, ASR systems designed for native speakers, have big problems when used by non-native speakers resulting in unacceptable recognition rates. Traditionally, speaker adaptation techniques are widely used to improve recognition rates of ASR systems. In this book the semi-supervised adaptation approach is also applied to improve the performance for non-native speakers. It is particularly important to have some kind of supervision in adaptation, if the ASR system is faced with high initial error rates, as is the case for the non-native speakers.

While achieving very good improvements on native as well as on non-native speech with the new, semi-supervised adaptation approach, it is believed that in the case of non-native speech there is room for more improvement. ASR systems usually expect words to be pronounced in a certain way.

If they are pronounced differently, which happens frequently in non-native speech, the automatic system is incapable of relating the 'wrong' pronunciation to the right word. The same problem arises with dialects. Solely applying speaker adaptation techniques is therefore believed to be not sufficient to achieve satisfying performance for non-native speakers but an additional modification of the pronunciation dictionary is necessary. The pronunciation adaptation method presented in this book introduces a new idea of generating non-native pronunciation variants. In contrast to known methods, this approach requires solely native speech to be successful.

The techniques developed in this research work will be tested on native and non-native speech. However, they are also applicable for dialect speech and are thus an important step towards the recognition of other than the 'average speech', which is an area of increasing interest.

This book is structured as follows: in Chapter 2, a general overview of ASR systems is given. Background information about the necessary pre-processing of the speech data and the theory of stochastic modelling of speech is given in Chapter 3 and Chapter 4, respectively. Chapter 5 describes the knowledge sources that are necessary in each ASR system. The improved speaker adaptation algorithm that was developed during this research work is described in detail in Chapter 6 together with the experiments and results that were achieved. After the applied approach for confidence modelling and its application to speaker adaptation is described in Chapter 7, the new pronunciation adaptation approach is presented in Chapter 8. A perspective of future work is given in Chapter 9 and the research is then summarised in Chapter 10.

2 ASR: An Overview

2.1 General Overview

Even though the process of communication is a very natural one and thus seems to be a fairly easy task for humans, there are several underlying operations that need to be carried out before the communication can be considered as being successful. In ASR all these operations that are handled by humans unconsciously are attempted to be fulfilled by machines or computers. Although some of the operations involved are well understood, the phenomenon of speech and language in general is still far from being clarified. The operations that were designed for ASR, try to model what we know about speech and language or what we assume about it. The models are often much simpler than reality and thus are imperfect.

The first operation in human communication comprises the 'hearing' of the message. We first have to realise that somebody is talking to us and then we have to listen to what he is saying. In the technically equivalent process this is done by recording the message via a microphone. The recording is either started manually or by speech detection algorithms. While humans can directly process the message as it is, the speech that is recorded with a microphone has to be converted into a machine-readable representation, as will be described in Chapter 3.

Both systems, the 'human' and the automatic one, need to have some knowledge about the sounds that are used. If one of the talkers uses sounds the other talker does not know, they cannot understand each other. This basically corresponds to two people speaking different languages. Also a common vocabulary that is a set of words known by the partners is needed.

When humans process the message they extract the meaning out of what was said. They can do so by inferring the meaning from the actual sequence of words that they recognised, because they can directly associate meanings to words and more important to word sequences.

Our technically equivalent system however, searches — due to the fact that we chose a stochastic modelling of speech — for the word or word sequence that was most likely spoken, given the acoustic signal. Even if this was successful, it is still far away from *understanding* what the meaning of this sequence of words is. Of course approaches already exist that try to extract the meaning out of recognised word sequences. Some of these systems are introduced in [Kom97, Sto00]. The automatic speech understanding (ASU) approaches are beyond the scope of this book and will not be considered any further here. But it is clear to see that in ASU several subsequent processing steps are needed, before a meaning can be extracted out of a spoken sequence of words. In Figure 2.1, a schematic overview of both speech recognition processes, the 'human' one on the right side and the automatic one on the left, is shown. The optional ASU part in the automatic system is shown with dashed lines. The fact that the 'human' process is shown as the processing of plenty of different information sources is supposed to indicate that these are very complex processes that are not completely clarified yet. More detailed investigations on the different human processes can be found in [Car86].

In this example, the speaker says 'What do you think about this?' The person he is talking to can easily understand that this was a question and he usually also knows what 'this' in this context refers to. Since the situation the speakers are in is a meeting in this example, it probably is a proposal the first speaker is drawing on a sheet of paper simultaneously. So the process of recognising and understanding is aided by several factors, such as the situation the talkers are in, the topic they are talking about, the expression on their faces, prosody and input other than speech, e.g. gestures. The listener can perceive all this at the same time while listening. They also possess something that is usually called 'knowledge of the world'. This terminology tries to express that humans have a certain knowledge about common things, that is acquired and extended during life. We know for example that a child has parents and that parents are a mother and a father, etc. In discussions we often do not really recognise every single word, but we guess a lot from the context, because we have all the additional information.

A topic that has long been neglected is that of using prosody as an information carrier in ASU and ASR systems, but some recent research has now addressed this problem, see [Kom97, Rey97, Kie96]. The automatic processes corresponding to those depicted in Figure 2.1 are explained in detail in Section 2.2.

A prototype system that tries to take into account factors like prosody, dialogue situation, topic, gestures, expressions on the speaker's face (emotion) and some (limited) knowledge of the world to *understand* the speaker's intention is the SMARTKOM system, see [Sma]. But although it is still not completely understood how the human brain processes speech, it seems that people are much better than the automatic systems when it comes to incorporating semantically relevant information in recognition.

In most ASR systems, however, and also in the system that is considered in this book, the knowledge of the world is even more limited compared to ASU systems and the only modality considered is the speech.

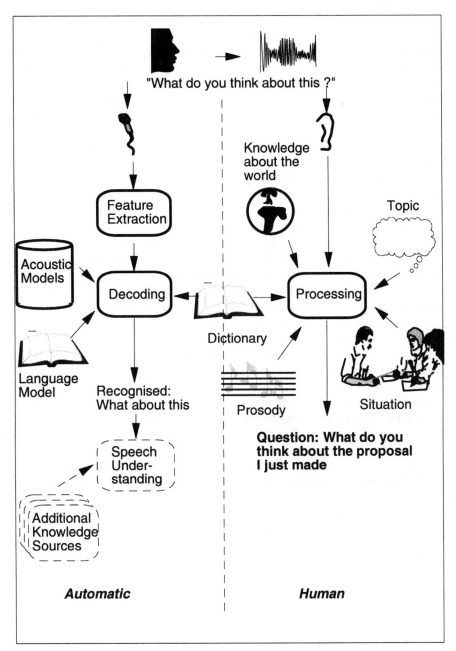

Fig. 2.1. Overview of a 'human' and an ASR system

2.2 Automatic Processing of Speech

In ASR several subsequent processing steps that are shown in Figure 2.1 are needed by the system, before the recognised word sequence is available. First of all, the speech signal as spoken by a speaker cannot directly be processed by computers. The first stage is thus to record the speech and to transform it into a machine readable format. Then the stage of finding the most probable word sequence follows and is typically called *search* or *decoding*. In this stage the algorithm uses information that is provided by the already mentioned different knowledge sources, which are the dictionary, the *language model* (LM) and the *acoustic models* (AMs). The dictionary contains the words that are known to the system together with their pronunciation. This helps the system to relate the spoken sounds to the actual words. The AM set provides a statistical model for each of the sounds of the speech that are used in the system. The LM gives a model for the structure of the language used. Both models will be described in detail in Chapter 5.

There are different types of ASR systems. If only single words are to be recognised, i.e. the words are spoken with long pauses in between, they are called *isolated word recognition systems*. This mode of operation is often used in command & control systems, where devices are controlled by speech. In this case often all the words in the dictionary are equally probable to be spoken at any time.

Systems that do not require pauses between the words and that allow complete sentences to be spoken, are called *continuous speech recognition* (CSR) systems or if the vocabulary is large, *large vocabulary continuous speech recognition* (LVCSR) systems. Examples for these kinds of systems are the widely known PC dictation systems. Here not all words are equally probable, but a LM is employed to predict the likelihood of word sequences. For LVCSR with the usual vocabulary sizes of 65,000 words this is especially necessary, because it would be computationally infeasible to find the correct sequence if all words were equally probable. A third type of ASR system is one that can process *spontaneous speech*. This kind of speech is encountered in conversations between humans. The way of speaking is very natural, thus often ungrammatical and including many filler words or hesitations, like 'uhm', 'ah', etc. Also speakers often stop within a sentence and repeat or correct parts of the sentence and so on. Processing spontaneous speech is the most complex task among the described ones and clearly needs several knowledge sources that help the understanding process.

For all of the above mentioned steps we need to find an appropriate description, model or knowledge source that we can use in our system and that tries to give a good view of the reality. It is necessary to find a good representation of the speech that can be easily processed by a computer, we need to find well adjusted AMs, a realistic way of describing how the different words we want to recognise can be pronounced and last but not least, we need to find a realistic model of the language i.e., well adjusted probabilities of the different possible word sequences. These various models need to be constructed before we can use them. Therefore, we need to distinguish between two different modes of our ASR system, the training and the testing mode. For the training of all our stochastic models we need large amounts of speech data that will henceforth be called *training data*. Speech data is used to train the acoustic models. It can also be used to derive the pronunciations for the dictionary but often the pronunciations are derived independently from an available dictionary. Statistical LMs are trained using large amounts of text data.

Once this is finished we can use the system and in order to test the performance of the models we determined during the training phase, we use *testing data* that should contain different speakers than the training data. If this was not the case we would tailor the system to the speakers that were contained in the training set. This might result in a very bad performance for new speakers. Using a lot of speakers during the training phase of the AMs will result in *speaker-independent* (SI) models. This means that the system is supposed to recognise all speakers well. The opposite is to tailor the system to one speaker using only speech data from this particular speaker during training resulting in a *speaker-dependent* (SD) system. While usually performing better for this speaker, the performance for other speakers will be worse. Often *speaker adaptation* techniques are used to close the gap between SI and SD systems. These techniques begin by employing SI models, thus initially providing a reasonable performance for all speakers. In the adaptation phase the AMs are modified so as to better recognise the current speaker. Then often the recognition performance comes close to SD systems but uses only a fraction of the speech data that would be needed for a complete SD training.

2.3 Evaluation of ASR Systems

In order to assess the performance of ASR systems, a measure needs to be defined that tells us how good our recognition system is. Well known measures are word accuracy (WA) and word error rate (WER). Before it is defined how these measures will be computed, the possible errors a recogniser can make will be listed first. Let's assume a CSR task and the spoken sentence was 'Which movies are being shown on TV tonight'. The best hypothesis found by the recogniser is 'Which news are shown on MTV at night'. Finding the minimal 'distance' between the two sentences by aligning them by *dynamic programming* (DP), see e.g. [Pau98], and comparing them, the errors made by the recogniser can be seen in Table 2.1.

The words 'Which', 'are', 'shown' and 'on' were recognised correctly. The words 'movies', 'TV' and 'tonight' were *substituted* by the words 'news', 'MTV' and 'night', respectively. 'being' was *deleted*, and 'at' was *inserted*.

Table 2.1. Correct and hypothesised sentence, aligned via DP, with S, D and I errors

Which	movies	are	being	shown	on	TV		tonight?
Which	news	are		shown	on	MTV	at	night?
-	S	-	D	-	-	S	I	S

So the types of errors a recogniser can make are *substitutions* (S), *insertions* (I) and *deletions* (D).

If N is the total number of words in the sentence, the word accuracy WA is defined as

$$WA = \frac{N - S - D - I}{N} * 100 \tag{2.1}$$

and the WER is

$$WER = 100 - WA = \left(1 - \frac{N - S - D - I}{N}\right) * 100$$

$$= \frac{S + D + I}{N} * 100 \tag{2.2}$$

Another frequently used term is the correct rate, which does not consider the insertions

$$corr = \frac{N - S - D}{N} * 100 \tag{2.3}$$

For the example shown in Table 2.1 we would get

$WA = \frac{8-3-1-1}{8} * 100 = \frac{3}{8} * 100 = 37.5\%$
$WER = \frac{3+1+1}{8} * 100 = \frac{5}{8} * 100 = 62.5\%$
$corr = \frac{8-3-1}{8} * 100 = \frac{1}{2} * 100 = 50\%.$

This means the WER is 62.5%. Throughout this book, mainly the WER will be used.

2.4 Adaptation in ASR Systems

As described in Section 2.1, at all processing stages we have to make assumptions to model reality. Often our models do not match reality, simply because the assumptions were wrong or the situation we modelled has changed in a real application. If this happens, the consequence will be that our system misrecognises more words or utterances. The greater is the difference between the situation and the assumption, the greater will be the number of misrecognitions. For instance, the AM or the LM could incorporate inappropriate assumptions. If the acoustic models were trained on clean speech, i.e.

speech that was recorded in a quiet environment and if the system is used in a public place with a lot of background noises, this will cause problems, because the models do not fit well anymore. If the LM was trained on read speech and now it is used in situations where spontaneous speech is mostly used, this will also be problematic, because filler words and hesitations that are typical in spontaneous speech were not contained in the LM training corpus.

Since it is impossible to account for all possible situations that might occur in real life, we must envisage a specific scenario when we construct an ASR system. But nevertheless, we have to be able to detect later that the situation changed and then be able to adapt to the new situation as much as possible. We try to adapt the models we are using to improve as much as possible the WA of our system in a certain situation.

It is the main concern of the research work presented here to adapt two of the models of an ASR system, the AMs (see Chapters 6 and 7) and the pronunciation dictionary (see Chapter 8). While the AMs are adapted to native as well as to non-native speakers, the pronunciation adaptation method is developed especially for non-native speakers.

3 Pre-processing of the Speech Data

In this chapter the necessary pre-processing of the analog speech waveform is briefly described. The description mainly follows [Pau98] and [Fur01].

First, by using a microphone, the speech signal or speech waveform is converted into an electrical signal. Figure 3.1 shows a block diagram of a typical speech analysis procedure, generating *Mel-frequency cepstral* coefficients (MFCCs). The different steps are explained in the following sections.

3.1 A/D Conversion

The electrical signal is band limited by a low pass (LP) filter and converted from an analog to a digital signal by an analog/digital (A/D) converter. Optionally the signal can be pre-emphasised. A/D conversion comprises sampling, quantisation and coding. To represent the analog signal as a sequence of discrete numbers, it is sampled. The sampling frequency needs to satisfy the Nyquist criterion, meaning that the sampling frequency f_s needs to be more than two times as high as the highest frequency component present in the function (f_b), in order to be able to fully reconstruct the analog signal:

$$f_s > 2f_b \tag{3.1}$$

During *quantisation* the waveform values are approximately represented by one of a finite set of values. Higher resolutions result in smaller quantisation errors but also in more complex representations. *Coding* assigns an actual number to each of these values. The sampling frequency typically ranges from 8 to 16 kHz with resolutions ranging between 8 and 16 bits. Optionally a DC offset that might be caused by the A/D converter can be removed.

3.2 Windowing

From the digitised signal so called 'analysis intervals' are extracted, which are the basis for the feature vectors that are to be computed from the signal. The size of these intervals typically ranges from 10 to 50 ms. The assumption behind this is that the shape of the vocal tract will not change in this period

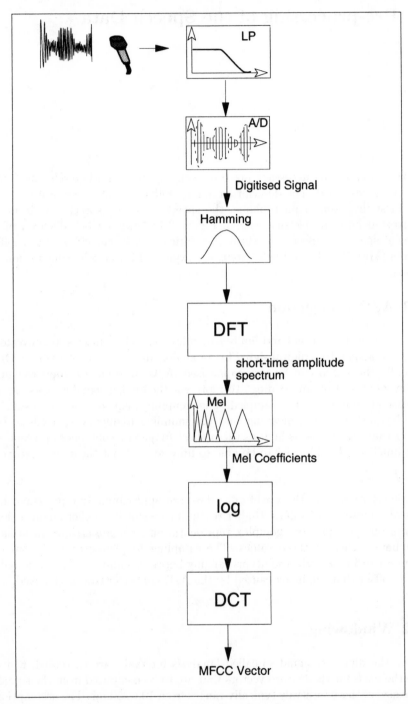

Fig. 3.1. Block diagram of the speech analysis procedure

of time and that therefore the signal can be considered to be quasi-stationary. Even though this is not true for some sounds, e.g., plosives, the transformation of the speech signal into a sequence of feature vectors based on such window sizes has proven to be useful in the past. In order to attenuate discontinuities at the window edges, an additional window function is applied to each extracted frame. The window function has to satisfy two characteristics, which are the high frequency resolution and the high attenuation of the side lobes. These requirements are contrary to each other. While the simple rectangular window has the highest frequency resolution, the attenuation of the side robes is too small. It is thus not suited to the analysis of signals with a large dynamic range. As a good compromise the Hamming window is often used:

$$w_H(n) = 0.54 - 0.46\cos(\frac{2n\pi}{N-1}) \tag{3.2}$$

where N is the number of samples with $n = 0, \ldots, N-1$. A plot of a Hamming window is shown in Figure 3.2. Since the multiplication of the signal with a window function reduces the effective length of the analysis interval (see [Fur01]), the analysis window should be shifted such that it overlaps with the previous one to facilitate tracing the time varying spectra. The short-time analysis interval that is multiplied by a window function is called a frame and the length of the interval is called the frame length or window size, the frame

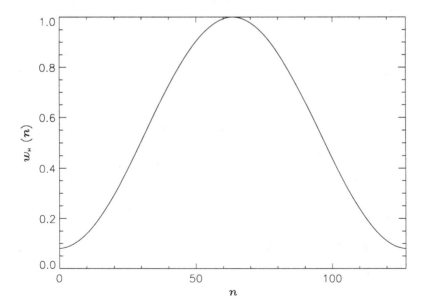

Fig. 3.2. Hamming window

shift interval is called frame interval or frame shift. The frame shift usually ranges from 5-20 ms.

3.3 Filter Bank Analysis

For analysing the speech spectrum, several methods, such as short-time auto-correlation, linear predictive coding or cepstral analysis are employed among others [Fur01]. The features that are computed from the speech signal in this research work are MFCCs. MFCCs are used in many ASR systems, because they offer good discrimination and the coefficients are not highly correlated. They are thus well suited to be modelled by probability density functions, see Chapter 4, that assume statistical independence.

In an effort to represent the auditory characteristics more precisely – the resolution of the human hearing system is different for different frequencies – the so called Mel-scale frequency axis was introduced. It corresponds to the auditory sensation of tone height. The relation between the frequency f in kHz and the Mel-scale Mel can be approximated as follows:

$$Mel(f) = 2592 log_{10}(1 + \frac{f}{700}) \tag{3.3}$$

To achieve this, usually a filter-bank is used that gives approximately equal resolution on the Mel-scale using overlapping, triangular filters as shown in Figure 3.3. Each DFT magnitude coefficient obtained is multiplied with the

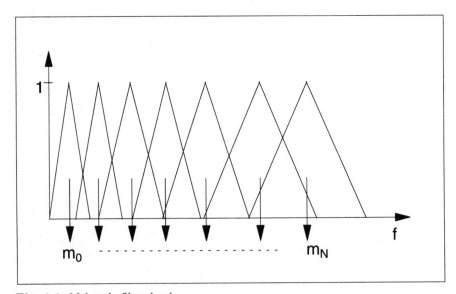

Fig. 3.3. Mel-scale filter bank

corresponding filter gain and the results are accumulated. This means that the resolution will be compressed more for higher frequencies than it will be for lower frequencies.

Then the logarithm is taken. If the digital cosine transform (DCT) of the log Mel-filter bank amplitudes m_j, see Figure 3.3, are calculated this yields the MFCC coefficients c_i.

$$c_i = \sqrt{\frac{2}{N_c}} \sum_{j=1}^{N_c} m_j \cos(\frac{\pi i}{N_c}(j - 0.5)) \tag{3.4}$$

where N_c is the number of filter-bank channels. To account for the time-varying features, the first and second order time derivatives are also computed and used. The settings for the pre-processing used in the experiments are described in Appendix A.

4 Stochastic Modelling of Speech

Hidden Markov modelling is a powerful technique, capable of robust modelling of speech. It first evolved in the early 70s. In the late 80s it became more and more used for speech recognition applications.

This technique considers speech a stochastic process and consists of modelling units of speech by Markov processes. Efficient algorithms exist for accurate estimation of *Hidden Markov Model* (HMM) parameters and they will be briefly described in Section 4.1. A detailed description of the underlying theory can be found in [Rab86, Rab89]. Using stochastic models to represent the words to be recognised, the basic problem in ASR is to find the word sequence $\hat{W}$ that has the highest *a posteriori probability* $p(W|X)$ among all possible word sequences given the speech signal $X(t)$:

$$\hat{W} = \arg\max_{W} p(W|X) \tag{4.1}$$

Instead of $X(t)$, $\mathbf{O}$ will be used to indicate that the sequence of feature vectors that is obtained from the speech signal $X(t)$ by the preprocessing procedure described in Chapter 3 is used.

Since the direct computation of the probabilities of all possible word sequences for maximisation is not tractable, Eq. 4.1 is transformed using *Bayes theorem*

$$p(W|\mathbf{O}) = \frac{p(\mathbf{O}|W)p(W)}{p(\mathbf{O})} \tag{4.2}$$

to

$$\hat{W} = \arg\max_{W} p(W|\mathbf{O}) = \arg\max_{W} \frac{p(\mathbf{O}|W)p(W)}{p(\mathbf{O})} \tag{4.3}$$

$p(\mathbf{O}|W)$ is the probability of the observation $\mathbf{O}$ given the word W and is determined by the AMs. $p(\mathbf{O})$ is the probability of the acoustic observation in general and since this is the same for all possible sequences, it can be omitted. $p(W)$ is the probability of the word sequence and is determined by the LM that will be described in more detail in Chapter 5.

In the following sections the basics of HMMs that are used for acoustic modelling are reviewed. The summary below does not represent an exhaustive discussion of the complete theory; rather it attempts to provide an overview of the basic HMM technology.

4.1 Hidden Markov Models (HMMs)

Hidden Markov Models consist of two stochastic processes, which enable the modelling not only of acoustic phenomena, but also to a certain extent of time scale distortions. A HMM system is capable of being in only a finite number of different states $s_1, \ldots, s_N$, each of them capable of generating different outputs. An example of a specialised HMM model is shown in Figure 4.1. At each time step $t, 0 < t < T$, a new state s_j is entered based on the transition probability a_{ij}. It gives the probability for moving from state s_i to state s_j and it is assumed that it depends only on the state s_i (this also includes self transitions, meaning transitions from state s_i to state s_i).

$$a_{ij} = P(q_{t+1} = s_j | q_t = s_i) \text{ with } \sum_{j=1}^{N} a_{ij} = 1 \qquad (4.4)$$

with q_t and q_{t+1} being the states at time t and $t + 1$, respectively, and N the total number of states. The transition probabilities are summarised in the matrix $\mathbf{A}$.

After a transition is made and a new state is entered, one out of K observation output symbols, v_k is emitted. The output is produced by another stochastic process b_j depending only on the current state s_j.

$$b_j(v_k) = P(o_t = v_k | q_t = s_j) \text{ with } \sum_{k=1}^{K} b_j(v_k) = 1 \qquad (4.5)$$

and $1 \le k \le K$. These output probabilities are summarised in a $N \times K$ matrix $\mathbf{B}$. Since the set of output symbols is a finite one, the incoming feature vector has to be mapped to one of the existing outputs in the set. This done by a technique called *vector quantisation* [Pau98].

Furthermore an HMM is characterised by its initial state distribution

$$\pi_i = P(q_1 = i) \text{ with } 1 \le i \le N \qquad (4.6)$$

Usually it is assumed that

$$\pi_i = \begin{cases} 1 & \text{if } i = 1 \\ 0 & \text{if } i > 1 \end{cases} \qquad (4.7)$$

Note that Equations 4.4 and 4.5 assume that subsequent observations o_t and o_{t+1} are independent. However, in the case of speech this is usually not the case.

A compact notation for the complete parameter set of an HMM model λ that is often used is

$$\lambda = (\mathbf{A}, \mathbf{B}, \pi) \qquad (4.8)$$

This model set describes a probability measure $P(\mathbf{O}|\lambda)$ for $\mathbf{O}$, i.e. the probability that this model generated an observed feature sequence.

When determining the most probable word that gave rise to the observation, we only decide which models gave rise to the observation but we do not know the underlying state sequence $s_1, \ldots, s_n$. That is the reason why these Markov Models are called *Hidden* Markov Models. A frequently used HMM topology is the left–to–right model. In this model, as time increases the state index also increases (or stays the same), i.e., the states proceed from left to right, emitting an observation vector each time a state is entered. It can thus model signals whose properties change over time, as speech signals do. An example of a left–to–right HMM model, that will be explained in more detail later in this section, is shown in Figure 4.1. Here none of the states can be skipped and the state sequence $s_1 \rightarrow s_1 \rightarrow s_2 \rightarrow s_2 \rightarrow s_3$ generates the observation sequence $o_1 \ldots o_5$. As will be explained in Chapter 5, one HMM for each phoneme is used. A word thus consists of several concatenated HMMs. The states that are drawn with dashed lines in Figure 4.1, are states of potential previous or subsequent HMM models.

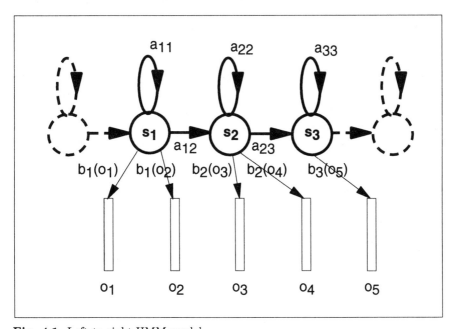

Fig. 4.1. Left-to-right HMM model

The fundamental property of all left-to-right HMMs is that the state transition probabilities have the property

$$a_{ij} = 0 \qquad \text{if } i > j \tag{4.9}$$

that is, no transitions are allowed to states whose indices are lower than the current state, which would in the context of speech correspond to transitions back in time.

Often additional constraints are imposed on the state transition coefficients to make sure that large changes in state indices do not occur; hence a constraint in the form

$$a_{ij} = 0 \qquad \text{if } j > i + \Delta \tag{4.10}$$

is imposed. Often Δ is set to two, i.e., no jumps or skips of more than one state are allowed, so that the model can repeat the current state or advance either one or two states. Rabiner et. al. [Rab93] found that recognition performance with this restricted model was superior to that of less constrained forms. Furthermore it would be infeasible to estimate the parameters of ergodic that is fully connected, HMMs. For the left-to-right model described, $\mathbf{A}$ is an upper triangular matrix and the bottom row, corresponding to the final state, is all zero, except for the final element, since there are no transitions away from the final state.

So far only the cases have been considered where the observations were discrete symbols chosen from a finite alphabet. Therefore, a discrete probability density could be used for each state and these models are called *Discrete Hidden Markov Models*. But the signals under consideration — speech signals — are continuous signals (or vectors). Even though vector quantisation techniques for these continuous signals exist to map them to discrete symbols, they might impose quantisation errors, which as a consequence may cause a degradation in recognition rates. Hence it is advantageous to use continuous observation densities. These HMMs are then called *Continuous Density Hidden Markov Models* (CDHMMs). Usually mixtures of normal distributions as defined in Equation 4.11 are used as the continuous output probability density functions (pdf) $b_j(o_t)$. As described in Chapter 3, the speech signal is transformed into a sequence of feature vectors $\mathbf{O} = o_1, \dots, o_N$. Given a speech vector o of length n, the probability density function of that vector being generated by mixture component k is:

$$
\begin{aligned}
b_{jk}(o_t) &= \mathcal{N}[o_t, \mu_{jk}, \mathbf{C}_{jk}] \\
&= \frac{1}{(2\pi)^{\frac{n}{2}} |\mathbf{C}_{jk}|^{\frac{1}{2}}} e^{-\frac{1}{2}(o_t - \mu_{jk})' \mathbf{C}_{jk}^{-1}(o_t - \mu_{jk})}
\end{aligned} \tag{4.11}
$$

The component densities are combined to give the state probability density:

$$b_j(o_t) = \sum_{k=1}^{K} c_{jk} b_{jk}(o_t) \tag{4.12}$$

where

$\mathcal{N}$	is the normal distribution
K	is the number of mixture components
c_{jk}	is the mixture weight of mixture component k in state j, $1 \le k \le K$
μ_{jk}	is mean vector associated with state j of mixture component k and
$\mathbf{C}_{jk}$	is the $n \times n$ covariance matrix associated with state j of mixture component k

Another class of HMMs is that of *semi-continuous* HMMs. In this case a pool of L density functions is used by all states. Their contributions to the pdf of each state is controlled by the mixture weight c_{jk} that is specific to each state.

4.2 Solving the Three HMM Problems

Given the HMM models that were described in the previous sections, there are three basic problems that need to be solved in order to employ the HMMs in real world applications:

- Given the observation sequence $\mathbf{O}(t)$ and the model $\lambda = (\pi, \mathbf{A}, \mathbf{B})$, how do we find the probability $P(\mathbf{O}|\lambda)$ of the observation given λ?
- Given the observation sequence $\mathbf{O}(t)$ and the model λ, how can a state sequence be determined that is in some sense optimal?
  The last two problems are solved during *recognition* or *decoding*.
- If we have a set of example observation sequences, how do we adjust the parameters of our models, so as to maximise $P(\mathbf{O}|\lambda)$?
  This problem is referred to as *training*.

4.2.1 Recognition

In order to solve the first problem we need to compute the probabilities for all possible state sequences of length T given the observation $\mathbf{O}(t) = o_1, o_2, \ldots, o_T$. An efficient way to do so is a method which is called the *forward procedure*. The forward variable α is the probability of being in state i at time t with the already observed feature sequence $o_1, \ldots, o_t$

$$\alpha_t(i) = P(o_1, \ldots, o_t, q_t = s_i | \lambda). \tag{4.13}$$

The solution for α can be obtained incrementally. First it can be calculated for the very first state

$$\alpha_1(i) = \pi_i b_i(o_1) \tag{4.14}$$

The subsequent values for α can then be computed recursively as follows:

$$\alpha_{t+1}(j) = [\sum_{i=1}^{N} \alpha_t(i)a_{ij}]b_j(o_{t+1}) \tag{4.15}$$

with $1 \leq t \leq T-1$ and $1 \leq j \leq N$. After this is done for all times up to $t = T-1$ the overall probability can finally be determined as

$$P(\mathbf{O}|\lambda) = \sum_{i=1}^{N} \alpha_T(i) \tag{4.16}$$

Since it is not tractable in many cases to compute the probabilities along each valid path, often a method called 'beam search' is used that follows only those paths whose differences in score w.r.t. the best hypothesis is below a certain threshold. If the difference is bigger, the corresponding hypothesis is not very likely to be the best one and can be discarded to save computation time.

In a similar manner the backward probability is determined assuming again that we are in state i at time t, this time provided that from this state on, the sequence $o_{t+1}, \ldots, o_T$ will be produced.

$$\beta_t(i) = P(o_{t+1}, \ldots, o_T, q_t = i|\lambda) \tag{4.17}$$

It is then defined that for $1 \leq i \leq N$

$$\beta_T(i) = 1 \tag{4.18}$$

Then $\beta_t(i)$ is with $t = T-1, \ldots, 1$ and $1 \leq i \leq N$.

$$\beta_t(i) = \sum_{j=1}^{N} a_{ij}b_j(o_{t+1})\beta_{t+1}(j) \tag{4.19}$$

In Section 4.2.3, the backward formula thus derived will be used together with the forward formula to solve the training problem.

4.2.2 Finding the Optimal State Sequence

For solving the second problem, an algorithm that is somewhat similar to the forward algorithm is used. Note however that there is no one single solution, since the solution for finding the optimal state sequence depends on the optimality criterion used. The algorithm used is called the *Viterbi algorithm*. It attempts to find the single best state sequence, $q = (q_1, q_2, \ldots, q_T)$ for the given observation sequence $\mathbf{O} = o_1, \ldots, o_T$. The difference compared to the forward procedure is that the sum over all possible state sequences is replaced by the maximum operator in order to determine the single best sequence. We define the quantity $\delta_t(i)$

$$\delta_t(i) = \max_{q_1,\ldots,q_{t-1}} P(q_1, q_2, \ldots, q_{t-1}, q_t = i, o_1, o_2, \ldots, o_t|\lambda) \tag{4.20}$$

By induction we get

$$\delta_{t+1}(j) = [\max_i \delta_t(i)a_{ij}]b_j(o_{t+1}) \qquad (4.21)$$

The initialisation is with $1 \leq j \leq N$

$$\delta_1(i) = \pi_i b_i(o_1) \qquad (4.22)$$

and

$$\psi_1(i) = 0 \qquad (4.23)$$

$\psi_t(j)$ is an array that we use to keep track of the argument that maximised Equation 4.21, because this is what we need to know to retrieve the optimal state sequence. Once the final state is reached we can use $\psi_t(j)$ to back trace the path through the trellis. A schematic view of a trellis is given in Figure 4.2. Because the Viterbi algorithm is faster and easier to implement, it is often used for decoding instead of the forward algorithm.

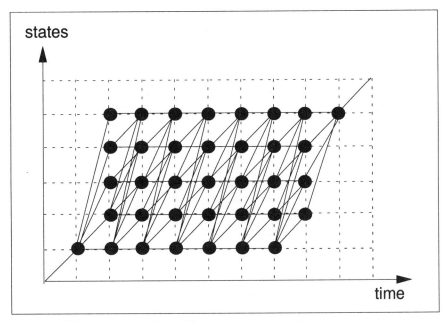

Fig. 4.2. Schematic view of a trellis

4.2.3 Training

The last and most complex problem is that of HMM parameter estimation, also called training. A large, pre-processed speech data base is needed, from which we can estimate the parameter set according to maximum likelihood

(ML) estimation. The estimation task is equivalent to finding those HMM parameters which maximise the likelihood of the HMMs having generated the training data. Since there is no known closed form solution to this problem, an iterative procedure is used. First initial values are assumed for $\mathbf{A}$, $\mathbf{B}$ and π. These estimates are then refined using the observations of the training data set. The parameters are then further refined and this process is iterated until no further improvement can be obtained. Here the Baum-Welch method [Bau72] is employed. It is identical to the Expectation-Maximisation (EM) algorithm for this particular problem. Let $\gamma(i,j|\mathbf{O},\lambda)$ be the expected number of transitions

from state q_i to q_j, given an observation sequence $\mathbf{O}$ and the model λ, and let $\gamma(i|\mathbf{O},\lambda)$ be the expected number of transitions from state q_i to any other state under the same conditions, then we can re-estimate the transition probability $\hat{a}_{ij}$ as follows:

$$\hat{a}_{ij} = \frac{\gamma(i,j|\mathbf{O},\lambda)}{\gamma(i|\mathbf{O},\lambda)} \tag{4.24}$$

where the hat indicates the new estimate and

$$\gamma(i|\mathbf{O},\lambda) = \sum_{j=1}^{N} \gamma(i,j|\mathbf{O},\lambda) \tag{4.25}$$

If $\gamma(i,j,t|\mathbf{O},\lambda)$ is the expected number of transitions from state q_i to q_j at time t, then

$$\gamma(i,j|\mathbf{O},\lambda) = \sum_{t=1}^{T-1} \gamma(i,j,t|\mathbf{O},\lambda) \tag{4.26}$$

$\gamma(i,j,t|\mathbf{O},\lambda)$ can be found with the aid of the forward-backward variables defined in Equations 4.15 and 4.19 in Section 4.2.1:

$$\gamma(i,j,t|\mathbf{O},\lambda) = \frac{\alpha_t(i)a_{ij}b_j(o_{t+1})\beta_{t+1}(j)}{p(\mathbf{O}|\lambda)} \tag{4.27}$$

and with 4.25 and 4.26

$$\gamma(i,j|\mathbf{O},\lambda) = \frac{\sum_{t=1}^{T-1}\alpha_t(i)a_{ij}b_j(o_{t+1})\beta_{t+1}(j)}{p(\mathbf{O}|\lambda)} \tag{4.28}$$

$$\gamma(i|\mathbf{O},\lambda) = \frac{\sum_{t=1}^{T-1}\alpha_t(i)\sum_{j=1}^{N}a_{ij}b_j(o_{t+1})\beta_{t+1}(j)}{p(\mathbf{O}|\lambda)} \tag{4.29}$$

by substitution from 4.19 we get:

$$\gamma(i|\mathbf{O},\lambda) = \frac{\sum_{t=1}^{T-1}\alpha_t(i)\beta_t(i)}{p(\mathbf{O}|\lambda)} \tag{4.30}$$

Correspondingly, the output probability density functions can be re-estimated with $\xi(j,k|\mathbf{O},\lambda)$ being the estimated number of times that state j

emits v_k, given the observation $\mathbf{O}$ and $\xi(j|\mathbf{O}, \lambda)$ being the estimated number of times that state j emits any symbol at all.

$$\hat{b}_{jk} = \frac{\xi(j, k|\mathbf{O}, \lambda)}{\xi(j|\mathbf{O}, \lambda)} \tag{4.31}$$

with

$$\xi(j, k|\mathbf{O}, \lambda) = \frac{\sum_{t:\mathbf{O}_t = v_k} \alpha_t(i)\beta_t(i)}{p(\mathbf{O}|\lambda)} \tag{4.32}$$

If the current model parameters are denoted $\lambda(\mathbf{A}, \mathbf{B}, \pi)$ and the re-estimated parameters $\lambda(\hat{\mathbf{A}}, \hat{\mathbf{B}}, \hat{\pi})$ Baum [Bau72], showed that

$$P(\mathbf{O}|\hat{\lambda}) \geq P(\mathbf{O}|\lambda) \tag{4.33}$$

meaning that the re-estimated model set is more or equally likely to have generated the observation compared to the current model set.

Since the procedure described involves many multiplications of small probabilities often the logarithmic values are used to avoid underflow.

5 Knowledge Bases of an ASR System

The main knowledge bases that are needed by an ASR system are the AMs, the pronunciation dictionary and the LM. These will be described in the following sections.

5.1 Acoustic Models

In Chapter 4 the principle of modelling units of speech as stochastic processes was introduced. One important question to be solved is that of which size of the unit to use. If the vocabulary of the application is small, as it is in a recognition system for isolated digits, then it is possible to use one HMM model for each word in the vocabulary, i.e. one for each digit. In this case samples for each of the words are needed for training. For larger vocabularies or CSR applications however, the number of different words is too large to be modelled in such a way. In these cases, the words are decomposed into sub-word units and one HMM model for each of these sub-word units is built using a training database that does not necessarily have to contain the words in the vocabulary. The sub-word units are then concatenated to build the words. Once models for these units are obtained the vocabulary can contain an arbitrary number of words, because all words can be built using the set of sub-word models. This is much more flexible than the whole-word modelling approach and needs no special training data when new words are added to the vocabulary. One commonly used subunit is the *phoneme*. In this research work, phones, which are physical realisations of phonemes are used, but syllables, demi-syllables or larger units can also be used, see [Rus92, Pfa97]. A phoneme is a minimal unit in the sound system of a language that causes a difference in the meaning of a word (see [Cry97, Lad93]). We use the SAMPA notation (see [SAM]) that is a computer-readable version of the international phonetic alphabet (IPA). The IPA aims at providing a notational standard for the phonetic representation of all languages (see [IPA]). An example for two German words that differ in one phoneme (and thus have different meanings) are the words 'Hüte' (Engl. hats) and 'Hütte' (Engl. hut). The phonemic transcriptions for these words are

Hüte   h y: t @
Hütte   h Y t @

For German we use a set of 43 symbols, a complete list of which can be found in Appendix C. Even when the same phoneme appears in different words, it is not necessarily pronounced in the same way, but its pronunciation is influenced by the phonemes that are pronounced prior to and after it. This effect is called co-articulation. To pronounce different phonemes, the articulators move to different places in the vocal tract. Now, while pronouncing the first phoneme, they are already 'on their way' to the next phoneme, so that the previous phoneme is influenced by subsequent phonemes. For example the /a/ in 'Hamburg' is pronounced different from the /a/ in 'Hallo' even though in the phonemic transcription the same phoneme /a/ is used:

Hamburg   h a m b U r k
Hallo   h a l o:

In practice, phonemes are often modelled in their left and/or right contexts, yielding *context-dependent* models that are called bi- or triphones, respectively. In the above mentioned example we would then distinguish between the triphones /h-a-m/ and /h-a-l/. If no context is taken into consideration, the models are called *context-independent* models or monophones.

The problem of co-articulation is more severe in continuous and spontaneous speech than it is in isolated word recognition. But in isolated word recognition applications it might also be helpful to account for these effects by using triphone models. The number of models then increases drastically. For German the number of triphones would be in theory $43^3 = 79507$. Even though not all these combinations exist in most of the languages due to phonotactic constraints, the number of triphones is usually still very high and thus requires a large amount of training data to estimate the models.

5.2 Pronunciation Dictionary

Another knowledge base of an ASR system is the pronunciation lexicon or pronunciation dictionary, henceforth briefly called dictionary. It contains information about which words are known to the system and also how these words are pronounced, i.e. what their phonetic representation looks like.

Figure 5.1 shows part of the dictionary that was used in the speaker adaptation experiments of Chapter 6. It gives the orthography for each of the words in the first column and its pronunciation in SAMPA notation in the second column. The pronunciations can be derived from existing dictionaries, such as the German Duden ([Dud]) or CELEX ([CEL]). They can be hand-crafted by experts or they can be generated automatically by grapheme-phoneme conversion tools. These tools try to generate the pronunciation of a

H	h a:
Hahnenweg	h a: n @ n v e: k
Halle	h a l e
Hamburg	h a m b U r k
Hamburger	h a m b U r g 6
Hannover	h a: n o: f 6
Heidelberg	h aI d @ l b E r k
Heierswall	h aI ? E r z v a l
Herion	h e: r j o: n
Hessen	h E s @ n
Hessischer	h E s I S 6
Hildesheim	h I l d E s h aI m
Hildesheimer	h I l d E s h aI m 6
Hilfetexte	h I l f @ t E k s t @
Hip Hop	h I p h O p
Hof	h o: f
...	...

Fig. 5.1. Excerpt from the pronunciation dictionary used

word, based on its orthography. In our case decision trees trained on CELEX are used for this purpose [Rap98]. The generated pronunciations are then manually checked by an expert.

Dictionaries of many ASR systems contain only the canonical or standard pronunciation that can be found in dictionaries like Duden or CELEX. The problem however with this standard pronunciation is that it is not always used in real life. This might be due to dialectal or accentual variation or caused by co-articulation effects in continuous and spontaneous speech. When the dictionary contains pronunciations that are seldom used by speakers, this will clearly cause problems later during recognition. If these inappropriate pronunciations are used to build the trellis, it will be hard for the recogniser to find the 'correct' word(s).

During training the dictionary also plays an important role. Usually the training data are transcribed automatically by *forced alignment*. That means the words that were spoken are known in advance and the recogniser is now forced to recognise this word sequence given the speech signal. To do so the pronunciation contained in the dictionary is used to find the phoneme boundaries in the speech signal. If inappropriate pronunciations are used, some speech segments will be assigned to the wrong phonemes and thus the parameter estimation for the corresponding phoneme models will not be optimal. A frequently occurring example is the following: In German many verbs end with the syllable /@ n/, like e.g. the word *gehen* (Engl. to go). Speakers often omit the /@/ and they say /g e: n/ instead of /g e: @ n/. Due to the fact that we use three-state HMMs with no skips, at least three frames (which is the fastest way of going through such a HMM) of the speech will be assigned to each of the phonemes. This example is depicted in Figure 5.2. So even if the /@/ was not spoken at all, there will be frames assigned to it and in this example these frames will in reality either belong to the /e:/ or the /n/. Thus for the /e:/ and the /n/ there will be frames missing, while for the /@/ wrong frames will be used for the parameter estimation. On the top of the speech waveform in Figure 5.2 the phoneme sequence that was actually spoken is shown. Below the phoneme sequence that was used for the alignment and the assigned portions of the speech are shown. The phoneme that was not pronounced (/@/) is shown with dotted lines. To avoid these kinds of problems, pronunciation variants like /g e: n/ should be included in the dictionary, so that both the canonical pronunciation and a variant can be used during training and testing.

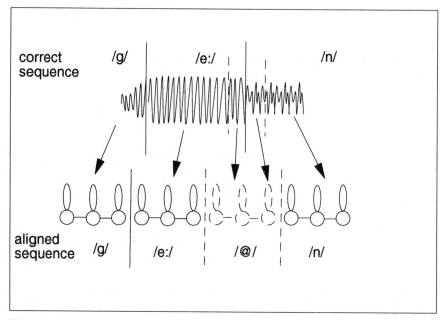

Fig. 5.2. Waveform of 'gehen' with spoken phoneme sequences and sequences used for the alignment

In this study the dictionary plays an important role since we later focus on speakers who speak with a foreign accent. In these cases the pronunciation used will deviate even more than the above example from what is considered the standard pronunciation. A more detailed description of the different techniques to generate pronunciation variants and the accompanying problems can be found in Chapter 8.

5.3 Language Models (LMs)

For small vocabulary isolated word recognition tasks, it is a valid approach to generate a trellis containing all possible words (and word sequences) and to search the best path through it. If longer phrases are to be recognised but the vocabulary is still relatively small, *finite-state grammars* are often used to define the possible sequences of words that can be recognised by the system. For continuous and spontaneous speech this is not at all tractable. Since languages usually follow semantic and syntactic constraints, these constraints should be used to help the recogniser during decoding. These constraints are captured in so-called LMs. The LMs are used to predict probabilities of certain word sequences. In Equation 4.2 in Chapter 4 the LM provides the probability for $p(W)$.

Usually large (written) text corpora (e.g. newspaper corpora) are used to estimate probabilities of word sequences based on their occurrence in the respective corpus. The probability of a word w depends on the words that were previously observed. The probability $P(W)$ of a word sequence W of length l can be written as

$$P(W) = P(w_1) \prod_{i=2}^{l} P(w_i | w_1, \ldots, w_{i-1}) \tag{5.1}$$

To keep the problem computationally tractable, the restriction for *n-gram* LMs is imposed that $P(W)$ depends only on the $n-1$ previous words. In most of the cases n is chosen to be two or three and the LMs are then called *bi-* and *tri-grams*, respectively. For word sequences that are unseen in the training data, a technique called *back-off* is used. If the word sequence $< w_1 w_2 w_3 >$ has not been observed, $P(w_3 | w_2, w_1)$ cannot be estimated directly. Instead, $P(w_3 | w_2)$ is used. If $< w_2 w_3 >$ has not been observed either, $P(w_3)$ is used; this is called the *uni-gram* probability. Often it is necessary to use a *word insertion penalty*. This penalises word transitions to avoid the recognition of too many short words instead of longer ones.

LMs are usually measured in terms of *perplexity*, describing their capability of predicting words in a text [dM98]. The perplexity gives the average word probability of a text as given by a LM that has typically been trained on other texts. In finite-state grammars, the perplexity can be explained as the average number of words that can follow another word. During recognition the LM probability is combined with the probability of the AMs using a LM weight. The general form of this is

$$P_{total} = P_{acoustic} P_{LM}^{\alpha} \tag{5.2}$$

or

$$logP_{total} = logP_{acoustic} + \alpha logP_{LM} \tag{5.3}$$

respectively. The higher the LM weight α the more the search will be restricted by the LM. It is necessary to weight the LM probability to account for inappropriate model assumptions. A more detailed description of language modelling can be found in [dM98].

6 Speaker Adaptation

Even though there exist various techniques to improve the recognition rates of SI systems, state–of–the–art SD systems still yield higher recognition rates than SI ones. If provided with the same amount of training data, they can achieve an average word error rate a factor of two or three lower than the SI system [Woo99]. But to train SD systems, large amounts of speaker specific speech data are needed and it is often not feasible to collect this data. Hence the use of speaker adaptation methods is appealing for solving this problem, since they promise to achieve SD performance, but require only a small fraction of speaker-specific data.

A second severe problem for today's SR technology is to deal with a mismatch between the training and the testing data. Often speech recordings are done in a special recording studio or at least special recording settings are fixed (e.g. the microphone used or the distance to the microphone). The HMMs are then trained with this data. But it is very difficult or sometimes impossible to have the same situation and/or settings when the real system (with these HMMs) is to be used. A good example is the case of mobile applications. Such applications are currently quickly evolving. As an extreme case, car applications can be considered, where there exist a vast variety of possible background noises. Here it is nearly impossible to cover all those background noises in the training data. This problem is present for the SI case as well as for the SD one, so in this example the use of SD models would probably yield better performance than SI models, but they still can't cope with the mismatch between the training and the testing conditions. Again, speaker and/or channel adaptation techniques are employed to weaken this mismatch either in the model- or in the feature space.

A third problem that can be solved by speaker adaptation techniques is that even though SI systems often exhibit a reasonable performance for most of the speakers, there are often speakers who are very badly recognised, even though human listeners would judge their voices as being normal and easily understandable. By adapting the system to these speakers, also for them reasonable recognition rates can be achieved.

In the following sections the different speaker adaptation techniques will be briefly reviewed before the new approach will be introduced. It is an important extension to the known method of MLLR adaptation and was developed to overcome the well-known problem of over-training on small amounts of adaptation data.

6.1 The State of the Art in Speaker Adaptation

To tackle the three above mentioned problems, speaker adaptation techniques are employed. They examine the speech from the new speaker and determine the differences between his way of speaking and the 'average' way of speaking, which is reflected by the SI models. Once these differences are known, either the SI models or the incoming features are modified, such that they better match the new speaker's acoustics. The more data that are available, the more specific differences can be determined. Often the incoming speech from the current speaker does not cover all sounds that are used by the system, so that a complete update of the model set is not possible. This is almost always the case if di- or triphones are used. Then the available data must be used to generalise, so that the models for those sounds that were not present in the adaptation data can also be modified.

Often the actual application and even more important the available amount of adaptation data determines which of the different adaptation methods to use. Throughout this book, the term *adaptation data* is used for the speech data of the speaker to whom we are trying to adapt our models. The general adaptation procedure is shown in Figure 6.1. The most common way is to adapt the HMM models, but also the features can be adapted. Feature-based approaches will be described in Section 6.1.1.

After an utterance is recognised the update functions that are used to adapt the HMM models are calculated, using the statistics that were extracted from the utterance and the recognition result, and then applied to the models. These adapted models are then used for recognition. This general procedure is valid for all of the methods that will be described in the following.

If large amounts of adaptation data are available from the new speaker in advance, the adaptation is termed *batch* adaptation. All the data are then processed in one step, the models are adapted in an offline way and then the user can start using the system with the models adapted especially to his voice. If no speech data are available beforehand the incoming data has to be used while the speaker is using the system. The data then consists of separate sentences or in case of a command & control application of single words or short phrases that were just spoken by the speaker. The models are adapted after each (set of) utterance(s) with very small amounts of adaptation data in an *incremental* way. The adapted models are then used to recognise the following utterance and so on. This method is referred to as incremental or *online* adaptation. The same procedure as shown in Figure 6.1 is applied. The difference now is that the update of the HMM models takes place after each adaptation step.

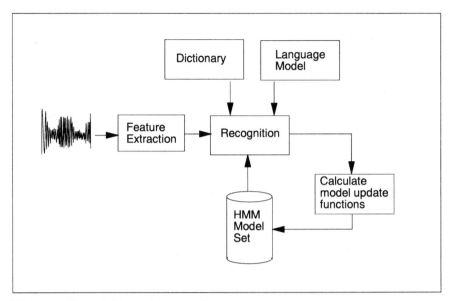

Fig. 6.1. General adaptation procedure

Further we distinguish between *supervised* and *unsupervised* adaptation approaches. In supervised adaptation the user has to speak predefined sentences, so that it is exactly known what was spoken. The recogniser can be forced to recognise those sentences by doing a forced alignment. We then have the possibility to directly compare how the sounds are modelled by the SI models and how the new speaker speaks them.

In contrast, in unsupervised adaptation, it is not known what was spoken, so a recogniser (using the SI models) needs to be used to first recognise the utterance. Then whatever was recognised is taken as the spoken words and based on this, the adaptation is conducted. The problem here is that it is not known if the utterance was recognised correctly. If it was misrecognised, the wrong sound units of the SI models will be compared with what was actually spoken. The differences between the two will be determined and the wrong sounds of the SI models will be adapted. If this happens repeatedly, the performance of our SI models will decrease instead of increase. In order to be able to deal with this problem, a semi-supervised adaptation strategy was developed, using CMs to decide which utterance to use for adaptation. In this way we do not need to force the user to speak some predefined sentences but we also try to avoid using misrecognised utterances for adaptation. The semi-supervised adaptation method is described in detail in Chapter 7.

Online, incremental and unsupervised adaptation with very small amounts of adaptation data can be considered to be the most challenging speaker adaptation task among all mentioned and this is the case that is going to be considered in the following. Since this research work concentrates on adaptation in the model space, the feature-based approaches are only briefly reviewed before the model-based approaches are explained in more detail in Section 6.2.

6.1.1 Feature-Based Approaches

Feature-based or feature space adaptation is also often termed *speaker normalisation* since it tries to compensate for the differences between speakers or environments. The differences between speakers mainly result from different vocal tract lengths, sizes of the mouth, or different nasal cavity shapes. The basic principle in feature-based approaches is to consider the features Y that were extracted from the speech data as being distorted compared to the original features X that were used to train the HMM models. The distorted features are mapped to the original ones using a transformation function:

$$X = F_\nu(Y) \tag{6.1}$$

Now the original HMM models can be used for decoding without modification. ν are the parameters to be estimated so as to maximise the likelihood of the adaptation data. The inter-speaker and environmental variability is here represented by a linear transformation in the spectral domain, which becomes a fixed shift in the log spectral domain. The simplest form of normalisation is now in widespread use and was introduced by Feng [Fen95]. It is called cepstral mean normalisation (CMN). It models a cepstral bias vector as the difference between the average cepstrum of the training and test speech on a sentence basis. This bias is then used to normalise the target speech prior to recognition. In contrast to the standard CMN approaches, where the cepstral bias is calculated as the mean vector of the cepstrum of the current utterance as a whole, the work of [Mor97] distinguishes between phonemes and uses only those that appear in the adaptation data for calculating the bias. More examples that also used more complex transformations can be found in [Neu95, Lee97, Cla98, Hwa97].

Another feature-based method is vocal tract length normalisation (VTLN). VTLN normalises the incoming data by using a linear frequency scaling factor or frequency warping. This is to account for the effect the vocal tract length has on the formant frequencies. Usually this vocal tract length scaling factor is chosen by testing different possible factors and choosing the one which maximises the performance for the incoming data [Pye97, Wel98]. Since only one parameter needs to be estimated, small amounts of adaptation data are sufficient, so that it can be applied on an utterance by utterance level during recognition. [Pye97, Wel98] showed that the use of VTLN eliminates the need to train gender dependent models, which is another method to improve the recognition rates of SI systems.

In general the feature-based approaches have been found to yield smaller improvements than model-based techniques [Neu95, San96]. However, it has been shown in [Pye97, Fen95] that a combination of both methods can be beneficial. E.g. [Pye97] showed that when VTLN is combined with MLLR adaptation the improvements are additive.

6.1.2 Model-Based Approaches

In contrast to the feature-based approaches, in model-based approaches the original models λ_x are mapped to the transformed models λ_y, which are supposed to better match the observed utterance, leaving the incoming features unchanged.

$$\lambda_y = G_\eta(\lambda_x) \tag{6.2}$$

The parameters to be estimated are η and they are again estimated so as to maximise the likelihood of the adaptation data.

Many different approaches exist for model-based speaker adaptation techniques. Two methods that are commonly established are MLLR and MAP (Maximum A Posteriori) adaptation and the one considered in more detail here is MLLR. While in MLLR adaptation parameters are pooled and updated with transformation matrices as will be described in the following section, the MAP approach conducts adaptation on the phoneme level. Although yielding a phoneme specific and thus finer adaptation than MLLR, the disadvantage of MAP is that it only updates the parameters of those phonemes that were observed in the adaptation data. As a consequence it needs a lot of adaptation data to reliably re-estimate the parameters for all phonemes used in the system, which takes rather long. Therefore MLLR was considered more appropriate, since an extremely fast adaptation with a very limited number of (short) utterances was required.

A new approach that was recently proposed by Kuhn et al. in [Kuh99, Kuh98b, Kuh98a, Ngu99b] is based on prior knowledge of speaker variation. Similar to the Eigenface approach in face recognition, the authors assume that there are certain characteristics that allow a distinction between speakers. For the purpose of speaker adaptation a SD model for each speaker is trained and then vectorised to build a speaker-specific super-vector. Dimensionality reduction techniques, like principal component analysis, are applied to keep only the most important parameters. If a new speaker is using the system, a vector that is a linear combination of these supervectors and as 'near' as possible to the new speaker is used to obtain an ML estimate for his parameters. So far this method only works for small-vocabulary tasks like e.g. alphabet recognition, but investigations were done by Botterweck [Bot00] to extend this technique to larger vocabularies. As we will see later in Section 8 the basic idea of this approach will become important for the pronunciation adaptation approach that is proposed by the author of this book.

6.2 Maximum Likelihood Linear Regression

Maximum Likelihood Linear Regression is a transformation-based approach.
It estimates a set of linear transformation matrices to modify the parameters
of the SI HMM models and was introduced by [Leg95b, Leg95c, Leg95a]. In
order to be able to calculate transformation matrices for smaller amounts of
adaptation data, the means are grouped into so called regression classes and
for each group one transformation matrix is calculated. If only a small amount
of adaptation data is available, one global transformation matrix is calculated.
If enough adaptation data are available, more than one transformation matrix
can be calculated. This clustering of the means makes it possible to update
parameters of more models than those that have actually been observed in
the adaptation data, since all means belonging to the same regression class
are always updated with the same matrix. As a consequence this approach
is quite fast, since it does not need to have examples for all means and
infrequently used phonemes can also be adapted.

The mean vectors of the Gaussian densities are updated using a $n \times (n+1)$
transformation matrix $\mathbf{W}$ calculated from the adaptation data by applying

$$\bar{\mu} = \mathbf{W}\hat{\mu} \qquad (6.3)$$

where $\bar{\mu}$ is the adapted mean vector and $\hat{\mu}$ is the extended mean vector:

$$\hat{\mu} = \begin{bmatrix} \omega \\ \mu_1 \\ . \\ . \\ . \\ \mu_n \end{bmatrix} \qquad (6.4)$$

ω is the offset term for regression ($\omega = 1$ for standard offset, $\omega = 0$ for no
offset). For the calculation of the transformation matrix $\mathbf{W}$, the objective
function to be maximised is:

$$F(O|\lambda) = \sum_{\theta \in \Theta} F(O, \theta|\lambda) \qquad (6.5)$$

By maximising the auxiliary function

$$Q(\lambda, \bar{\lambda}) = \sum_{\theta \in \Theta} F(O, \theta|\lambda) log(F(O, \theta|\bar{\lambda})) \qquad (6.6)$$

where

O	is a stochastic process, with the adaptation data being a series of T observations being generated by this process
λ	the current set of model parameters
$\bar{\lambda}$	the re-estimated set of model parameters
θ	the sequence of states to generate O and
Θ	all possible state sequences of length T

the objective function is also maximised. Iteratively calculating a new auxiliary function with refined model parameters will therefore iteratively maximise the objective function [Bau72] unless it is already a maximum. Maximising Equation 6.6 with respect to $\mathbf{W}$ yields

$$\sum_{t=1}^{T} \gamma_s(t) \mathbf{C}_s^{-1} \mathbf{o}_t \hat{\mu}_s' = \sum_{t=1}^{T} \gamma_s(t) \mathbf{C}_s^{-1} \mathbf{W} \hat{\mu}_s \hat{\mu}_s' \tag{6.7}$$

where

s is the current state
C_s is the covariance matrix of state s
o_t is the observed feature vector at time t and
γ_s is the total occupation probability of state s at time t

Since we have one (or l) regression matrice(s) shared by several Gaussians, the summation has to be performed over all of these Gaussians.

$$\sum_{t=1}^{T} \sum_{r=1}^{R} \gamma_{s_r}(t) \mathbf{C}_{s_r}^{-1} \mathbf{o}_t \hat{\mu}_{s_r}' = \sum_{t=1}^{T} \sum_{r=1}^{R} \gamma_{s_r}(t) \mathbf{C}_{s_r}^{-1} \mathbf{W}_l \hat{\mu}_{s_r} \hat{\mu}_{s_r}' \tag{6.8}$$

where

R is the number of Gaussians sharing one regression matrix.

Eq. 6.8 can be rewritten as

$$\sum_{t=1}^{T} \sum_{r=1}^{R} \gamma_{s_r}(t) \mathbf{C}_{s_r}^{-1} \mathbf{o}_t \hat{\mu}_t' = \sum_{r=1}^{R} \mathbf{V}_r \mathbf{W}_l \mathbf{D}_r \tag{6.9}$$

where $\mathbf{V}_r$ is the state distribution inverse covariance matrix of Gaussian r scaled by the state occupation probability:

$$\mathbf{V}_r = \sum_{t=1}^{T} \gamma_{s_r}(t) \mathbf{C}_{s_r}^{-1} \tag{6.10}$$

and $\mathbf{D}_r$ is the outer product of the extended means:

$$\mathbf{D}_r = \hat{\mu}_{s_r} \hat{\mu}_{s_r}' \tag{6.11}$$

The left hand side of Equation 6.8 is further denoted as $\mathbf{Z}_l$, which is an $n \times (n+1)$ matrix for regression class l. If the individual matrix elements of $\mathbf{V}_r$, $\mathbf{D}_r$ and $\mathbf{W}_l$ are denoted by $v_{r,ij}$, $d_{r,ij}$ and $w_{l,ij}$, respectively, we get after further substitutions

$$z_{l,ij} = \sum_{q=1}^{n+1} w_{l,iq} g_{l,i,jq} \tag{6.12}$$

where $g_{l,i,jq}$ are the elements of a matrix $\mathbf{G}_{l,i}$, the dimension of which is $(n+1)(n+1)$.

The $z_{l,ij}$ and $g_{l,i,jq}$ can be computed from the observation vector and the model parameters:

$$w'_{l,i} = \mathbf{G}_{l,i}^{-1} z'_{l,i} \qquad (6.13)$$

$w_{l,i}$ and $z_{l,i}$ are the ith rows of $\mathbf{W}_l$ and $\mathbf{Z}_l$, respectively, and so $\mathbf{W}_l$ can be calculated on a row by row basis. This set of linear equations is solved by Singular Value Decomposition (SVD).

Since in our implementation the Viterbi algorithm is used for decoding, each speech frame is assigned exactly to one distribution, so that

$$\gamma_{s_r}(t) = \begin{cases} 1 & \text{if } \theta_t \text{ is assigned to state distribution } s_r \\ 0 & \text{otherwise} \end{cases} \qquad (6.14)$$

Since we follow only the best path and thus $\gamma_{s_r}(t) = 1$, Eq. 6.8 becomes

$$\sum_{t=1}^{T} \sum_{r=1}^{R} \mathbf{C}_{s_r}^{-1} o_t \hat{\mu}'_{s_r} = \sum_{t=1}^{T} \sum_{r=1}^{R} \mathbf{C}_{s_r}^{-1} \mathbf{W}_l \hat{\mu}_{s_r} \hat{\mu}'_{s_r} \qquad (6.15)$$

In the case of mixture distributions, where the output distributions are made up of M mixture components, Eq. (6.8) can be easily extended to include all the mixtures and their corresponding weight. However, in the current studies only the maximum distribution is used. A detailed discussion of this matter can be found in [Leg95b].

There are different ways of assigning the Gaussians to different regression classes, as for example described in [Gal96a]. The regression classes can be based on broad phonetic classes, so that all Gaussians belonging to a phonetic class will share the same transformation matrix. Another possibility is to compute the distance between the Gaussians and cluster them based on these distances. For the experiments described the Euclidean distance measure was used for the assignment of the Gaussians to regression classes. Since for the applications considered here, mainly the use of one regression class is interesting, an extensive description of this matter is omitted and for the major experiments only one global regression class was used.

The covariance matrices used were diagonal ones and only the means were adapted. Several studies showed that an adaptation of the variances yields only minor or no improvements, see [Gal96b].

6.2.1 MLLR for Small Amounts of Adaptation Data

In [Leg95a], Leggetter and his co-authors describe MLLR as a fast adaptation approach, but they also state that still several sentences of adaptation data are necessary in order to reliably calculate the transformation matrices $\mathbf{W}_l$.

Some approaches exist that try to deal with the insufficient data problem in speaker adaptation. Many of them combine MLLR and MAP adaptation to conduct both in one adaptation step. [Fis99] combines both approaches such that the 'old' mean in MAP adaptation is replaced by the MLLR transformed mean. Standard MLLR with one regression class was used. In this paper it was confirmed again that a certain number of sentences is necessary for successfully conducting MLLR.

Another approach that was used by [The97, Ngu99a], computes the MAP estimates for those Gaussians that were observed. For the unobserved Gaussians the MLLR estimates are used. Also [Dig95a] combines transformation-based adaptation with MAP, such that first all parameters are updated by applying the transformation and then conducting MAP adaptation for the observed parameters. Other approaches that try to solve this sparse data problem for the Bayesian or MAP approach are [Neu95, Dig95b, Rot99, Shi97, Chi97, Zav96, Mat98]. The main reason to use MLLR in combination with MAP however was to overcome the problem, that MAP conducts a phoneme specific adaptation and thus needs even more adaptation than MLLR does in order to improve the models. In this way they could cope with smaller amounts of adaptation data compared to using MAP alone but still MLLR needed several sentences of adaptation data.

The problem in MLLR adaptation is that using only single words or short phrases as the basis for the transformation matrix calculation may cause problems, since non-representative data would be used. The derived matrices are likely to be erroneous, which might cause numeric problems for the matrix inversion in Eq. 6.13 and lead to unreliable adaptation of the means. Experiments showed exactly this to be the case. They are described in Section 6.2.4. However, for many applications it is important to adapt very quickly, even if only a few single words or utterances are available.

An approach that tries overcome this problem for MLLR is that of Chou and Chesta [Cho99, Che99]. They use a method called *Maximum A Posteriori Linear Regression* (MAPLR). This technique does not use ML estimation for the computation of the transformation matrices but Bayesian estimation, thus imposing constraints on the transformation parameters. By not relying solely on the adaptation data for the estimation of the transformation parameters, the authors intend to avoid the usage of malformed transformation parameters. The parameters η are no longer assumed to be fixed as it is done for ML estimation but they are considered to be random values that are described by their probability density function $p(\eta)$ which is derived from the SI models. The authors state that the complexity of the MAPLR formulation can be reduced to be the same as MLLR. However, no results for this case are available yet. Siohan [Sio00] extends MAPLR by introducing a tree-structure for the prior densities of the transformations yielding *Structural Maximum A Posteriori Linear Regression* (SMAPLR). In this way the regression classes can be dynamically chosen depending on the amount of adaptation data. Each node in the tree is associated with a transformation and the transformation of a node is used to update all parameters of all child nodes. If not enough data are available, the transformation of the parent node is used. Transformations can also be propagated to the child nodes and be combined with the adaptation data belonging to that node.

Byrne and Gunawardana [Byr99] use a moment interpolation variant of the EM algorithm that optimises a discounted likelihood criterion to cope with sparse adaptation data. The approach is therefore called *Discounted Likelihood Linear Regression* (DLLR). They show that the EM algorithm can be viewed as alternating minimisation between a parameter set Θ determined by the models and their parametrisation, and a family of desired distributions $\mathcal{D}$. The alternating minimisation is done between the parameter set and the desired distribution. The moment interpolation algorithm replaces the forward procedure of this alternating minimisation, such that it interpolates between the previous desired distribution and the projection of the previous parameter set to get the next desired distribution. The authors state that the success of MLLR in using the moment decay depends strongly on how the desired distribution is initialised.

6.2.2 The Weighted MLLR Approach

For the kind of speech data that was used (the speech corpora are described in more detail in Appendix A) and that is typical for command & control applications, the standard MLLR method is not suitable if used as described above, since the adaptation data consists of only single words or short utterances. While the MAPLR approach relies on an appropriate choice of the prior densities, which is not a trivial task and the choice of priors strongly influences performance in Bayesian estimation [Huo96, Huo95], a comparable problem arises for the DLLR method, which depends heavily on the initialisation of the desired distribution. The new method that is presented is an important contribution to MLLR adaptation because of its simplicity and capability to reduce the WERs on very small amounts of adaptation data. It introduces a weight for the calculation of the adapted mean, taking into account both the adaptation data and the SI models, without the need to choose any prior densities. It is capable of solving the problem of insufficient data. A description of this approach can also be found in [Gor99].

Using a Static Weight. Applying the standard MLLR approach yielded an increase in word error error rate on our test data. Obviously the small amounts of adaptation data lead to unreliable estimates of the transformation matrices. In order to solve this problem, the adapted mean was not any longer calculated as described in Eq. 6.3, but some prior knowledge was additionally taken into account. The prior knowledge was represented by the SI models and the final adapted mean was now calculated as follows:

$$\hat{\mu}_{j,l}^k = \alpha_l \bar{\mu}_{j,l}^k + (1 - \alpha_l)\hat{\mu}_{j,l}^{k-1} \tag{6.16}$$

where α_l is a weighting factor, controlling the influence of the 'standard MLLR-adapted' mean $\bar{\mu}^k_{j,l}$ (according to Eq. 6.3), k is the current utterance and $(k-1)$ the previous one, respectively. $\hat{\mu}^{k-1}_{j,l}$ is the mean of Gaussian j estimated from the previous utterance $(k-1)$ and l is the regression class index. That means that the adapted mean is the weighted sum of the 'old' and 'new' mean and in this online approach, this results in an interpolation of the original value and the one estimated from the current utterance. The influence of the current utterance can be adjusted by the weighting factor α_l. This weighting was inspired by MAP adaptation. α_l is kept constant for all of the processed utterances, which means that each utterance of the new speaker has the same influence on the models. Standard MLLR as described in [Leg95a] is a special case of the weighted approach. Setting α_l to one would yield exactly Eq. 6.3.

If the system is used for a longer time, it could be beneficial not to weight each utterance equally but to account for all previously processed utterances with a stronger weight. A weighting scheme that achieves this is described in the following section.

Using a Dynamic Weight. Usually major changes need to be made to the models whenever a speaker or environment change takes place to quickly adapt the SI models to the respective condition. But if the speaker is using the system for a longer time in the same environment, no big changes but rather a fine tuning is required. So once this broad adaptation to the speaker and the environment is done, additional utterances should not bring about major changes any more. To achieve this kind of fine tuning on larger amounts of data, the adaptation should be weighted with the number of speech frames that have already been observed from that specific speaker. This can be achieved by calculating α_l as follows:

$$\alpha^k_l = \frac{n^k_l}{\tau^{k-1}_l + n^k_l} \tag{6.17}$$

and

$$(1 - \alpha^k_l) = \frac{\tau^{k-1}_l}{\tau^{k-1}_l + n^k_l} \tag{6.18}$$

where n^k_l is the number of frames that were so far observed in the regression class l and τ^{k-1}_l is a weighting factor, the initial value of which will be determined heuristically. The influence of the adaptation that was already done in the past and thus the resulting models gradually increases, whereas the influence of current utterance decreases. Now Eq. 6.16 can be rewritten as

$$\hat{\mu}^k_j = \frac{\tau^{k-1}_l \mu^{k-1}_j + n^k_l \bar{\mu}^k_j}{\tau^{k-1}_l + n^k_l} \tag{6.19}$$

where τ^k_l increases by n^k_l after each adaptation step

$$\tau_l^k = \tau_l^{k-1} + n_l^k. \tag{6.20}$$

Using this weighting scheme, α_l^k and thus the influence of the most recently observed means decreases over time and the parameters will move towards an optimum for that speaker.

The advantage of the dynamic weighting compared to the static one is that if after a longer adaptation process a series of misrecognition occurs, e.g. due to some unexpected noise, the models are only slightly modified and this should not have a big effect on the performance. In contrast, when weighted statically the utterances including the unexpected noise would strongly influence the models, which could cause a sudden performance decrease.

6.2.3 Implementation Issues

The general procedure of calculating the matrices and updating the means is described in Figure 6.2. After an utterance is recognised (Viterbi), the optimal state sequence is determined. That means that for each state of the best path the corresponding Gaussian is determined. The means of the Gaussians are duplicated, then also including a regression offset in their 0th parameter. For each regression class, the relevant statistics (feature vector, means, covariance matrix, see Section 6.2) are extracted and accumulated. After the minimum number of frames for adaptation is reached, one **G** matrix is calculated for each row of the transformation matrix for that regression class. This is done using the SVD algorithm. As stated in [Leg95c], this method is suggested, since it is capable of dealing with ill-conditioned matrices. The **G** matrices may sometimes be ill-conditioned due to the lack of data. After the full transformation matrix **W** has been calculated, the means of the Gaussians are adapted according to Equation 6.16 or 6.19, respectively. These models are then used for recognising the following utterance(s) and so on.

Perform Viterbi recognition			
Determine optimal sequence of states and Gaussians			
Duplicate means of the Gaussians			
FOR each regression class			
	IF adaptation was conducted in previous iteration		
	THEN	Initialize adaptation parameters	
	ELSE	Use parameters calculated in previous adaptation iteration	
	FOR each speech frame in current utterance		
		Accumulate z, g, frame counter += 1	
		IF Minimum number of frames reached	
		THEN	Calculate transformation matrix W, frame counter = 0
Update all means			
Copy back all updated means			

Fig. 6.2. General MLLR procedure

Computational Effort. The computational requirements depend on the number of regression classes, number of Gaussians and the minimum number of frames used for adaptation. For each element $g_{l,i,jq}$ 3 multiplications are necessary and there are $(n+1) \times (n+1)$ elements per $\mathbf{G}_{l,i}$ matrix. For each row of the transformation matrix there is a separate $\mathbf{G}$ matrix that means if we have L regression classes and $\mathbf{W}$ is an $n \times (n+1)$ matrix, we have to calculate $n \times L$ $\mathbf{G}$ matrices and perform SVD $n \times L$ times. Then all the M means are updated with a vector-matrix multiplication making $n \times (n+1)$ multiplications. So in total the computational requirement is $3 \times (n+1)^2 + M \times n \times (n+1)$ multiplications and $n \times L$ singular value decompositions.

Memory Requirements. Memory for storing the following values is needed

z matrix:	$m \times n \times rc \times 4bytes$
G matrix:	$rc \times n \times n \times m \times 4bytes$
transformation matrix W_l:	$m \times n \times rc \times 4bytes$
extended means:	number of Gaussians $\times m \times 4bytes$
updated means:	number of Gaussians $\times n \times 4bytes$

6.2.4 Experiments and Results

The algorithm was tested using the database described in Appendix A. The corpora used for testing were the following:

- Addresses: German addresses and city names ($\approx$ 23 utterances/speaker, dictionary size = 130)
- Commands: German commands for controlling all kinds of consumer devices ($\approx$ 234 utterances/speaker, dictionary size = 527)

All adaptation experiments were conducted in online, unsupervised mode, as described in Section 6.1. The recogniser was restarted for each speaker and each corpus, using the SI models. Both weighting schemes were tested using a different number of frames for calculating the transformation matrix $\mathbf{W}_l$. This was achieved by specifying a minimum number of frames. Whenever this minimum was reached, adaptation was conducted at the end of the utterance. Using a minimum number of frames of one thus corresponds to conducting adaptation after each utterance. When 1000 frames were used, it was conducted after approximately four to five utterances in the case of the described database. When using the static weight α^1, it was kept constant for all utterances in one experiment. All following figures show the results in % WER.

First the overall results are shown in Table 6.1. They used a minimum number of 1000 frames and those values for α and τ that performed best

---

[1] Since mostly one regression class is considered, the index l is dropped when talking about the weighting factor

(this was the case for $\alpha = 0.4$ and $\tau = 1000$ for the addresses and $\alpha = 0.1$ and $\tau = 1500$ for the commands). Also the results for the standard MLLR approach are shown, which corresponds to setting $\alpha = 1$. In this case the WER increases drastically by 65.6% and 22.3% to 10.1% and 15.9% for the address and command corpus, respectively. That means the standard approach fails for the small amounts of adaptation data considered. In contrast to this, both weighted approaches can improve the WER on the command corpus by 20.7% for the static and 23.8% for the dynamic weight. Even on the address corpus impressive improvements of 9% can be achieved. This is remarkable, considering that it consists of 23 utterances per speaker only and that adaptation is thus conducted only four or five times on the average.

Table 6.1. WERs for the different adaptation methods

Method	Addresses	Commands
SI	6.4	13
Standard MLLR	10.1	15.9
Static Weight	5.8	10.3
Dynamic Weight	5.8	9.9

However, the two weighting factors need further examination. The results when varying α are shown in Figures 6.3 and 6.4 if one and 1000 frames were used for adaptation, respectively. The figures show both corpora (addresses and commands, indicated by 'addr' and 'comm') together with the results for the SI system (indicated by 'addr(SI)' and 'comm(SI)').

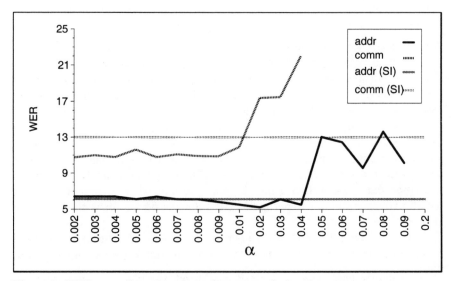

Fig. 6.3. WERs as a function of α, adaptation after each utterance

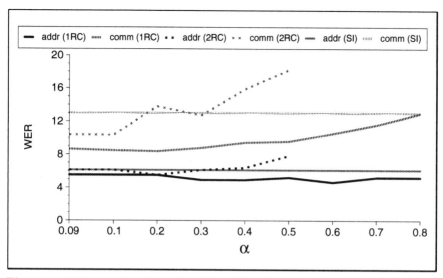

Fig. 6.4. WERs as a function of α, adaptation after 4-5 utterances

It can be seen that if one frame or one utterance is used, the performance of the adaptation depends very much on the choice of α. While clearly outperforming the SI system for values beneath $\alpha = 0.01$ on the command corpus, a sudden drastic increase in WER can be observed for larger values of α. For the address corpus the weighted MLLR is only better than the SI system for a few values of α and for $\alpha > 0.04$ an increase in WER can be observed. Also the WER can increase to more than 90%, which can be seen in more detail in Table B.1 in Appendix B. In the Figure these values are not plotted since they exceed the range of the y-axis. No results for two regression classes are shown because only very few values for α were found that improved the WER and even then the improvements were rather modest.

If a minimum number of 1000 frames (corresponding to four or five utterances) is used as shown in Figure 6.4 (the use of one or two regression classes is indicated by '1RC' and '2RC'), the performance depends also on α to a high extent, however the WER increase is not as drastic as in the one frame case and if one regression class is used, the SI system is always outperformed. If two regression classes are used, the WER varies between 10% and 20% for the commands, respectively, and this means that it is worse than the SI system for many values of α. These results again underline, that the closer α gets to one, the worse is the performance. Remember that setting $\alpha = 1$ corresponds to the standard MLLR approach.

In general the results show that using 1000 frames rather than one frame is better. The reason is that more data are available to calculate the adaptation statistics which makes the estimation of the transformation matrix $\mathbf{W}_1$ more reliable. For the same reason bigger improvements can be observed for the command corpus than for the address corpus. More utterances are available per speaker and thus adaptation can be conducted more often. The results for two regression classes are better than the SI results but worse than for one regression class. Using two regression classes can be considered as a kind of 'reduction' of adaptation data, since it takes more utterances to e.g. collect 1000 frames for the respective classes, resulting in a 'delayed' update of the models. Thus for the small amounts of adaptation data considered, the reduced number of model updates results in smaller improvements.

To test the performance of the approach under various conditions, we used SI models that were trained on different databases or under different conditions than the test set. The results for these experiments are listed in Table B.2 in Appendix B. The approach shows consistent improvements in all cases considered, however always involving a big empirical effort for finding the optimal α. Please note that these experiments include non-native speakers and for them also big improvements can be achieved. More focus will be put on non-native speakers later in Chapter 8.

The sensitivity of the static approach w.r.t. the weighting factor α makes it risky to use in real applications or changing experimental conditions. When carefully chosen, drastic improvements can be achieved. However, if an inappropriate value is chosen, a drastic performance decrease might also happen. Also the dependency on various factors such as number of frames, vocabulary and number of regression classes is not desired. An offline tuning of α would be necessary. In a real applications all these factors are subject to frequent changes, which renders the tuning of α impossible and underlines the need for a more robust approach.

The dynamic weight is expected to overcome these problems and the following figures summarise the results when using the dynamic weight. Figures 6.5 and 6.6 show the results for the command corpus, using 1, 1000, 2000, 5000 and 10000 frames and one and two regression classes, respectively. The number of frames was increased since it could be seen in the previous experiments that an increase yielded better improvements. For the address corpus no results are shown, since it is too small for conducting experiments using such a large number of frames. As can also been seen in Table 6.1, the improvements are slightly better compared to the static weight. Improvements of up to 23.8% can be achieved using one regression class and 1000 frames. Again, using a minimum number of one frame is not very stable and an increase in WER can be observed. However the maximum WER is 18% as compared to more than 90% in the static case. For all other cases that used more than one utterance, the WER remains more or less constant, even if τ is varied in a very wide range. This is also true if two regression classes are used, except for using one frame. This is again problematic and very high WERs (up to 80%, see Table B.3 in Appendix B exceeding the range of the y-axis) can be observed for different initial values of τ. Furthermore, results for two regression classes are not as good as the results for one regression

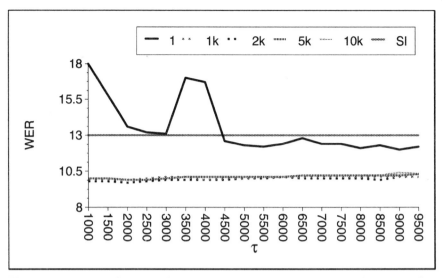

Fig. 6.5. WER as a function of τ, one regression class, commands, adaptation after each utterance, using a different number of frames

class but the difference between one and two classes is not as big as in the static case. In general however, varying the initial value τ does not really harm performance and even if a non-optimal τ is chosen the performance varies only slightly between the different initial weights. This is in contrast to static weight, where the choice of an inappropriate α even produced results that were worse than the SI system.

These results indicate, that the dynamic weight is clearly superior to the static weight and it is therefore chosen for all subsequent experiments that involve MLLR adaptation.

It was expected that the performance further increases when increasing the amount of data for calculating **W** by increasing the number of frames to 2000, 5000 and 10000 frames. But as the figures show, this increase in adaptation data does not have a big impact on the WERs. Using 10000 frames even causes the performance to decrease again. So providing more data for the calculation of **W** does not always help. This was also confirmed by [Wit99b], who found that after some 'saturation' point is reached in MLLR adaptation, adding more adaptation data does not help to get further improvements. A possible explanation is that the parameter estimation is always to some extent erroneous, since non-representative data are used. If the matrices are estimated on too large portions, the re-estimation process might be confused. For the future experiments that means that one regression class and a minimum number of frames of 1000 will be used.

To underline the superiority of the dynamic weight, the change of one arbitrarily chosen component of a mean vector is shown in Figure 6.7 after

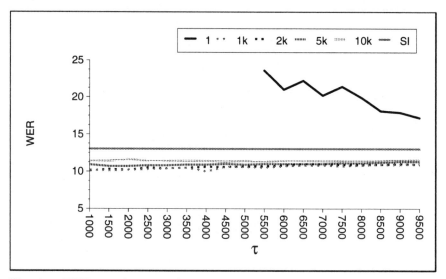

Fig. 6.6. WER as a function of τ, two regression classes, commands, adaptation after 4-5 utterances, using a different number of frames

each adaptation step, for the static and the dynamic weight, respectively. The change is defined as the difference between the current and the previous value of that mean component. While the change for the static weight is rather big even after many adaptation steps, it slowly decreases if the dynamic weight is used. In this example adaptation was conducted on the command corpus after every 1000 speech frames and the means were updated 88 times in total.

This shows that the longer the dynamic weighting is used, the smaller the changes of the means will be and only a kind of a fine tuning is done. This is reasonable, since in that way a possible series of misrecognitions will not really harm the adapted models as it would do in the static case.

6.3 Summary

In this chapter the weighted MLLR approach was presented. It was shown that it outperforms the widely used standard MLLR approach by far since it is capable of operating successfully on very small amounts of adaptation data. The standard MLLR approach when tested on the same data drastically decreased performance. The new approach is very fast and shows significant improvements already after a very small number of short utterances that were spoken in isolation. For the standard MLLR approach several continuously spoken sentences are necessary to improve the SI performance. Woodland [Woo99] states, that the proposed approach is one way to overcome the problem of small amounts of adaptation data for MLLR adaptation.

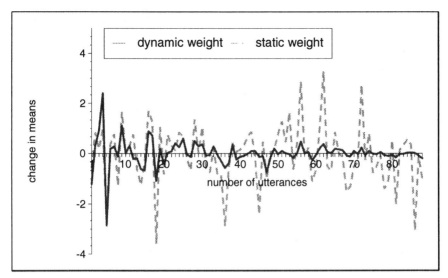

Fig. 6.7. Change of one arbitrarily chosen mean component

Two weighting schemes, a static and a dynamic one, were introduced. Both can greatly improve the WERs and yield comparable results, though the dynamic one is slightly better but, more important, it is also much more stable and insensitive to the actual choice of the weight. In the dynamic case only an initial value needs to be chosen that is then updated taking into account the number of speech frames observed. Even a strong variation in the initial value of τ could not really harm the overall performance, while slight modifications of the static weight α sometimes led to a deterioration of the results. This clearly indicates that the dynamic weight is much more robust and thus preferable for any kind of application.

The use of one global regression class outperformed two and more regression classes, since for the small amounts of adaptation data used any further splitting of data delays the update of the respective parameters, resulting in a performance that is worse.

The best results were achieved by using 1000 frames, corresponding to ten seconds of speech, to estimate the transformation matrix $\mathbf{W}$. As a consequence of the experiments described in this chapter, for all following experiments, one regression class and a minimum number of 1000 frames will be used.

The adaptation approach presented in this chapter considered unsupervised adaptation only. On one hand it is known that the use of unsupervised adaptation might harm performance if too many misrecognitions occur and are used for updating the model parameters. On the other hand, an enrolment phase in which supervised adaptation is conducted to adapt to a new speaker, is not acceptable for most of the applications. To prevent an increase in WER if the adaptation is faced with high initial error rates, CMs are employed to guide adaptation. This topic will be explained in detail in Chapter 7.

7 Confidence Measures

The majority of ASR systems achieve recognition rates that are well below those achieved by humans. Especially with increasing vocabulary size, the probability that words are confusable increases, thus making it more difficult to recognise the correct word. Furthermore, many commercially important speech recognition tasks require the ability to understand spontaneous rather than isolated speech, which is an even bigger problem. Although this provides a user friendly user interface, it poses a number of additional problems, such as the handling of out of vocabulary (OOV) words, disfluencies and acoustical mismatch. And unaware of the technology limitations, users expect the system to work properly, even if their utterance includes hesitations, false starts and sounds like uhm's and ah's.

In order to keep the error rates low, so called CMs are used. These CMs try to judge how reliably a word or utterance was recognised by the system. If it is considered to be unreliable, which means the probability that it was misrecognised is high, it is rejected. If the probability that it was recognised correctly is high, it is accepted and processed as usual. In many applications, the cost of processing a misrecognised command far exceeds the cost of rejection. A rejected utterance can be followed by a re-prompt to the user, asking him to repeat his utterance.

Furthermore, the use of CMs can be highly useful in unsupervised speaker adaptation schemes (see [Ana98, Fen95]). As described in Chapter 6, the unsupervised adaptation algorithm has to rely on the recogniser to deliver the correct word. If this is not the case, the wrong HMM models are modified. Using CMs, only those utterances that were previously accepted by the CM are used for adaptation.

In the literature, many approaches to the problem of estimating confidence for a recognition result have been presented. In the following the most important ones are briefly reviewed and our approach, which uses a set of features, is presented. We introduce some new features that are related to phoneme durations. Their combination with traditional features can improve the classification rates. As a classifier, a neural network (NN) is used. It will be shown that this approach is also capable of simultaneously identifying OOV words. This approach is then combined with weighted MLLR speaker adaptation.

7.1 The State of the Art in Confidence Measures

The calculation of CMs can either follow the *one-step* or the *two-step* strategy. In the one-step strategy, the decoding criterion is modified, see e.g. [Jun97, Eve01], such that the decoded string is that one, which obtains the highest confidence score w.r.t. all possible string hypotheses. Here the confidence test is not applied to a single string, but to all possible strings. In addition, a special training can be applied to maximise the discriminability between the HMM models, see [Lle96a, Lle96b, Koo98, Koo97].

In the two-step scheme, a CM for the hypothesised word is computed in a post-processing step, i.e. after recognition is completed. This approach is more flexible than the first one, since it leaves the search unchanged. The way the CM is computed can be varied more easily. This is an important point, since the most effective CM often depends on the settings and recognition task. On the other hand a disadvantage is that the post-processor that calculates the CM has to rely on a probably error prone segmentation that was obtained by the recogniser. However, it seems that more attention so far has been dedicated to the two-step procedure and several different kinds of two-step schemes for CMs have been evaluated in the past. They comprise

- The use of explicit alternate, anti-keyword, garbage, junk or filler models together with a log likelihood ratio (LLR) test
- The use of a set of features that are collected during recognition or extracted from the recogniser output, which could e.g. be a N-best list or a word-graph. A decision whether to accept the word/utterance or not is either made by simple thresholding if only very few features are used or by combining features using some kind of classifier, if multiple features are used.

In general, Williams [Wil98b] states that it is relatively easy to identify non-speech events or noises, because they lead to gross model mismatches and will thus lead to large reductions in confidence. In contrast to that, an OOV word that differs maybe in only one single phoneme from a known word, will reduce the confidence only slightly. Therefore a bad pronunciation lexicon may have an adverse effect on CM performance. This suggests that a careful modelling of pronunciation is necessary, especially in conjunction with CMs that measure reliability on the phoneme level. In particular non-native speakers will not only use different phonemes but also different phoneme sequences than expected. These phenomena would result in very low assigned confidence scores. A more detailed discussion about the necessity and importance of adequately modelling pronunciations for non-native speakers is given in Chapter 8. Please note that apart from the approaches described above, there is a variety of other approaches. The use of filler models or features are only the most widely used ones and are described in more detail in the following section. Different approaches can be found in [Rue97, Mar97, Uhr97, Gun98, Wei95, Wes98, Riv95, Jit98, Bar98, CG95, Jun98, Bou99, LC99].

7.1.1 Statistical Hypothesis Testing

The approaches that are described in this section consider the problem of computing the CM as a statistical hypothesis testing problem. The decoded hypothesis is tested against the so-called alternate hypothesis. Traditionally there are two kinds of likelihoods that are used for the alternate hypothesis. One is derived from *anti-keyword* models, the other one from *filler models*

For small vocabulary tasks, often anti-keywords are used (other terms such as alternate, junk or garbage models are also widely used in the literature, however in this document only the expression anti-keywords will be used. Additionally we will talk about decoded words, but of course also decoded phones, sentences or utterances could be used). One possibility is to train an anti-keyword for each word in the vocabulary, using all the available speech data that do not represent the word under consideration or using a set of words that are considered very confusable w.r.t. this word. Special discriminative training methods can be used for training the anti-keyword models and/or keyword models to increase the discriminability between them. This could be either minimum verification error (MVE) training as it is used in [Mod97, Suk97, Set96, Suk96] or minimum classification error (MCE) training as e.g. used in [Mod97, Suk97, Suk96, Set96].

The anti-keywords can also be grouped, so that one anti-keyword covers all non-speech events and another one covers all OOV words as done in [Rah97, Riv96, Jia98, Kaw98b]. These are then often termed filler models and use context independent phoneme models.

When using an anti-keyword for each of the words in the vocabulary, the verification of the decoded word is carried out by testing the *null hypothesis* H_0 versus the *alternate hypothesis H_1*, in which the null hypothesis represents the hypothesis that a specific word really exists in the segment of speech (and was therefore decoded correctly) and the alternate hypothesis that the decoded word does not exist in the segment of speech under consideration (and was therefore decoded incorrectly). The likelihood ratio (LR) is designed to determine whether or not a sequence of feature vectors Y was generated by a given set of probability densities, defining the following test (which is the *Neyman-Pearson-Lemma*)

$$LR = \frac{P(Y|\lambda^c)}{P(Y|\lambda^a)} \geq \zeta \tag{7.1}$$

where H_0 means Y was generated by keyword model λ^c and H_1 that Y was generated by alternate model λ^a, ζ is a decision threshold (or sometimes also called operating point) and λ^c and λ^a are HMMs corresponding to the correct and alternative hypothesis, respectively. If the probability for the alternate hypothesis is bigger than the one for the null hypothesis, the LR will be very small and lie under the previously defined threshold, so that the word will be rejected. In practice, the alternate hypothesis can be considered as a means to reduce the sources of variability on the CM. If both the null and alternate hypothesis are similarly affected by some systematic variation in the observation, these effects should be removed by calculating the LR. Implementations of this approach are [Kaw97, Kaw98b, Kaw98a, Mod97, Gup98, Ram98, Wu98, GM99, Che98, Wil98d]. Also [Ros98] uses acoustic LR-based CMs, but he enhanced the n-gram LM with them by adding a variable for the encoded acoustic score to each state in the LM. This was to prevent an inserted or substituted word in the recognition output from causing an additional error that would occur if the LM was used without the additional acoustic confidence score.

7.1.2 Using a Set of Features

It is common practice to formulate CMs by collecting a set of features previous to or during recognition and to decide according to the values of these features, whether to accept the utterance or not. The decision can be made by simple thresholding if only one or a few features are used, as done in [Jia98, Wil98a, Tor95, Ida98, Riv96, Ber97], or by using some kind of classifier, if several features are used. This classifier transforms the set of features into a measure representing the confidence for that word. One important step in the construction of such a classifier is the choice of features. These features need to be correlated to the occurrence of misrecognitions. One often-used feature is the length of a word, since it is known that very short words are misrecognised more often than longer words. Another feature that is often used, is the difference in scores between the first- and second-best hypothesis in a N-best list or word graph. The recogniser is assumed to be 'sure' about its own recognition result, if the score of the first-best hypothesis is much bigger than the score of the second-best hypothesis. If the scores are very close, the probability that the words are rather similar and that thus one or more words were confused with others is usually higher. In this thesis the feature-based approach was adopted and some new features were derived and combined with traditionally used features. The whole feature set is described in more detail in Section 7.4. The selection of features strongly depends on whether the recognition system is an isolated word or CSR system. If a word graph is computed, more features can be employed, such as word graph density or the number of occurrences of a certain word in the same place in the word graph. Similarly, in a CSR system, LM scores can be taken into account, which have proven to be a very reliable indicator of confidence in [Sch97]. Comparable to [Sch97], [Zep97] also varied the weights between acoustic score and LM to obtain different hypotheses, counted the number of times each word appeared in these hypotheses and used this as a feature.

CMs can be also used for re-ordering N-best lists or word graphs, as proposed by [Eid95, Cam97, Ber98, Fet96, Lin98].

Apart from the NN classifier, Bayesian classifiers [Cox96], linear discriminant analysis (LDA) classifiers [Sch97, Dol98, Pao98, Wei97], or binary decision tree classifiers [Cha96, Net97, Eid95], have also been used.

Neural Net Classifier. A classifier that is widely used and that is also adopted in the approach described here, is the NN classifier [Ben98, Sch97, Kem97, Kem99, Col95]. One big advantage of the NN classifier is that it can realise a general non-linear function if the NN has at least one hidden layer. This is advantageous, because not all features one might select depend linearly on the fact that a word is misrecognised. A good example is the speaking rate. It is a good indicator for recognition errors since for very slow and very fast speaking rates, many recognition errors occur. When calculating the correlation of the speaking rate with the WER, the result will be a very small correlation. The reason is not that there is no relation between the two, but that it is not simply linear (but quadratic). So for combining several features, including non-linear ones, the use of a NN seems beneficial. Sometimes it is also helpful to use the logarithm of a value (e.g. number of phonemes of the utterance) to achieve a higher correlation. Before the application of the NN classifier is explained in detail, a very brief introduction to NNs in general will be given in the following section.

7.2 Neural Networks

The following description is intended to be a very brief introduction to NNs. For a more detailed overview please refer to [Hop85, Rit92]. NNs try to model the human brain, where a vast number of small processing units, called neurons, highly inter-connected via so-called synapses, work in parallel. This is in contrast to the way computers work nowadays, because these often have only one (or a few) processing units that process the tasks sequentially.

Just like the human brain, a NN consists of many cells (which correspond to the neurons and are called nodes) that are connected via direct links and which are working in parallel. Information is processed by transferring activation patterns from one (group of) neurons to another (group of) neurons through special weighted connections.

Each node is characterised by its inputs, its activation function and its outputs. All the incoming information is summed up, taking into account the weights associated with each link and the activation function is applied to it. Depending on the resulting output value, the node is active and 'fires' or not. The decision when the node is active is based on a threshold. The weights associated with the links can also have negative values. Then the activation of the following neuron will be decreased instead of increased. In order to use such a net to solve a specific problem, it has to learn its task in a training phase. For a classification problem, we need to provide the net with a set of example patterns together with the desired output (i.e. the correct result). This is called supervised learning. With the help of standard learning algorithms, the weights of the links are adjusted, so as to achieve the smallest possible classification error. The nodes are usually organised in layers, and the resulting net is called a *multi-layer perceptron* (MLP). The nodes of the first layer are called input nodes and the nodes of the last layer are called output nodes. All the nodes of the layers that might be in between the in- and output layer are called hidden nodes and the layers are also called hidden layers. Only feed-forward nets are considered here. In feed-forward nets only links to subsequent layers exist. Figure 7.1 shows a NN with three layers, seven input nodes for the features $f_1 - f_7$, four hidden nodes and two output nodes, o_1, o_2. All nodes of the input layer are connected with all nodes of the hidden layer and all nodes of the hidden are connected with all nodes of the output layer.

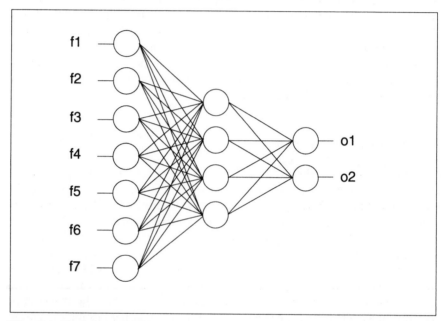

Fig. 7.1. Feed-forward NN

7.2.1 Activation Function

The activation function of a node is computed from the output of preceding units, usually multiplied with the weights connecting these predecessor units, the old activation function and its bias:

$$act_j(t+1) = f_{act}(net_j(t), act_j(t), \kappa_j) \tag{7.2}$$

where

act_j is the activation function of unit j in step t
net_j is the net input to unit j in step t
κ_j is the threshold (bias) of unit j

The activation function of the input and output layer was in the experiments always set to the *sigmoid*-function[1].

$$f_{act}(x) = \frac{1}{1 + e^{-x}} \tag{7.3}$$

The net input $net_j(t)$ is computed as

$$net_j(t) = \sum_i wgt_{ij} out_i(t) \tag{7.4}$$

This yields the activation function

$$act_j(t+1) = \frac{1}{1 + e^{-(\sum_i wgt_{ij} out_i(t) - \kappa_j)}} \tag{7.5}$$

where

i	is the index of the predecessor node
wgt_{ij}	is the weight of the link from unit i to unit j
$out_i(t)$	is the output of unit i in step t

The *sigmoid*-function is shown in Figure 7.2.

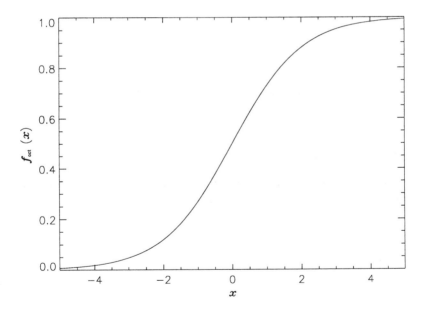

Fig. 7.2. *sigmoid* activation function

---

[1] SNNS was used to build our nets, [Ins98]

For the hidden nodes, the hyperbolic tangent (*tanh*) activation function (also called *symmetric sigmoid*) was also used, which is shown in Figure 7.3:

$$act_j(t+1) = tanh(net_j(t) + \kappa_j) \qquad (7.6)$$

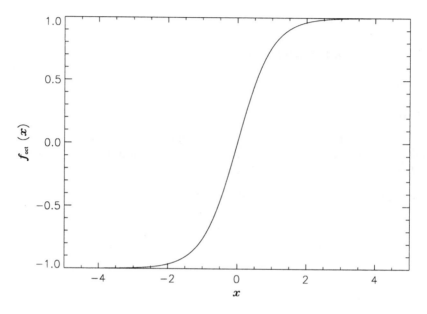

Fig. 7.3. *tanh* activation function

7.2.2 Output Function

The output function computes the output of the current node from its activation function. In our experiments the identity function was used

$$out_j(t) = f_{out}(act_j(t)) = act_j(t) \qquad (7.7)$$

7.2.3 Learning in NNs

Once the topology of the NN, i.e. the number of layers and nodes in each layer, is defined, the most important question is how to adjust the weights. This learning is often based on the Hebbian Rule, which says that a link between two nodes is strengthened if both nodes are active at the same time. The general form of this rule is:

$$\Delta wgt_{ij} = g(act_j(t), teach_j)h(out_i(t), wgt_{ij}) \qquad (7.8)$$

with

teach$_j$ teaching input (desired output) of node j
g(...)   function, depending on the activation of the unit and
           the teaching input
h(...)   function, depending on the output of the preceding
           node and the current weight of the link.

In supervised learning the net is provided with the training pattern files together with the desired output. An input pattern is presented to the network and is then propagated through the net until it reaches the output layer. This is called the forward propagation phase. The output from this layer is then compared to the desired output. The difference between the actual output out_j and the desired output $teach_j$ is used together with the output out_i of the source node to compute the necessary changes of the link weight wgt_{ij}. For the nodes for which no desired output is available, which is the case for the hidden and input nodes, the Δs of the following layer that have already been computed, are used. This phase is called backward propagation. The goal is to minimise this error during training and is called *gradient descent* method. Two learning algorithms were used, *Vanilla Backpropagation*, which is the one described above and *Backpropagation Chunk* [Ins98].

The definition for Vanilla Backpropagation is as follows

$$\Delta wgt_{ij} = \varrho \delta_j out_i \tag{7.9}$$

$$\delta_j = \begin{cases} f'(net_j)(teach_j - out_j) & \text{if unit } j \text{ is an output node} \\ f'(net_j) \sum_k \delta_k wgt_{jk} & \text{if unit } j \text{ is a hidden node} \end{cases} \quad \delta_j \text{ is the}$$

difference between the output of unit j and its teaching output $teach_j$, ϱ is a (constant) learning factor and k is the index of the successor to the current unit. The weights are updated after each training pattern, i.e. after each forward and backward pass. In the Backpropagation chunk algorithm the weights are updated after a *chunk* of training patterns was processed. In our case, 50 patterns were used.

As a stopping criterion for training an evaluation set was used and the mean square error (MSE) was computed, which is the summed, squared difference between the teaching and the actual output, divided by number of training patterns n, on this evaluation set.

$$MSE = \frac{1}{n} \sum_j (teach_j - out_j)^2 \tag{7.10}$$

Usually for the first iterations a decrease in MSE can be observed up to a certain point from which it again increases. So the training was stopped when the MSE of the previous iteration was smaller. The optimal learning parameter (specifying the step width of the gradient descent during training), was different for each net. They are in the range between 0.1 and 1.0 and were varied in 0.02 steps. The second learning parameter was the maximum allowed difference between teaching and output value of the output nodes during training. Here 0.1 turned out to be most useful one. Detailed statistics about the patterns used for training and testing can be found in Appendix A.

7.3 Evaluating Confidence Measures

When constructing classifiers, it is necessary to test their performance before they can be used. Several metrics to do so exist in the literature. The ones used in this book are described in the following. In a classification task two types of errors might occur, the so-called *Type I* and *Type II* errors. A Type I error occurs when the null hypothesis H_0 is rejected even if it was true and Type II errors are those that consider the alternate hypothesis H_1 as correct when it is incorrect. Type I errors are therefore also called *false rejections* and Type II errors *false alarms*. The sum of Type I and Type II errors divided by the total number of tests is called *unconditional error rate*. It is clear to see, that the unconditional error rate strongly depends on the 'difficulty' of the task. A trivial classifier that simply accepts all recognised words, achieves an unconditional error rate that is equal to the recogniser error rate. It is thus much easier to achieve a low unconditional error rate for a small vocabulary, isolated word recognition task, for which the recogniser achieves high accuracies, than for a LVCSR task, where the baseline performance is much lower. In general, we are interested in the correct classification rate (CC) that means how many samples have been classified correctly. The goal when constructing a classifier is to accept as many as possible correctly recognised utterances and reject as many as possible misrecognised utterances. The constructed classifier should be better than a trivial classifier. The performance of our CM was measured in terms of classification error rate (CER), which is the number of incorrectly classified patterns divided by the number of total patterns. Also the correct and false alarm rates (C_a and F_a, respectively) and correct and false rejection rates (C_r and F_r, respectively) are listed. The C_a rate directly corresponds to the number of utterances that were recognised correctly in the overall system. False alarms are those utterances that have been classified as being correctly recognised although they were misrecognised, so this corresponds to a recognition error of the baseline system.

7.4 CM Features

In addition to some known features, new ones are introduced in this research work. The complete feature set is described in the following sections. A description of our approach can also be found in [Gor00a, Gor00b].

7.4.1 Phone-Duration Based Features

It can be observed that when a word is misrecognised, there is a mismatch between the segmentation (and thus the phoneme durations) found by the recogniser and the durations that can be found in the training data. Often, when the speech signal is displayed together with the phoneme boundaries that were found by the recogniser, misrecognitions can be identified just by looking at the signal. In extreme cases, boundaries are e.g. found right in the middle of a (real) vowel. These kinds of errors should be detected in any case. This motivated the use of some new features that are related to phoneme durations. To the knowledge of the author, these kinds of features have so far not been used by anybody else. During the training of the HMM models, the smoothed distributions of durations of all phones are determined, based on a forced alignment of the training data. The phoneme durations that are later found by the recogniser are compared to the durations that were determined on the training data. Since it is well known that the speaking rate strongly influences the phoneme durations, the speaking rate was estimated and the found durations were normalised accordingly. The speaking rate α_c was estimated as follows, see [Kom97]:

$$\alpha_c = \frac{1}{N} \sum_{i=1}^{N} \frac{dur_i}{\bar{x}_p} \tag{7.11}$$

where N denotes the number of phones in the observed utterance and/or in the past few utterances, dur_i is the duration of the i-th phone segment (recognised as phone p) in the utterance and $\bar{x}_p$ is the mean length of the corresponding phone p learned during training.

Some of the features were multiply used, e.g. not normalised, normalised by number of frames, normalised by acoustic score of the best hypothesis or normalised by the speaking rate.

The feature set comprises the following features:

1. *n_toolong01*, *n_toolong05*: Number of phoneme durations in the best hypothesis W_1 that are longer than the 1% and 5% percentile $\eta_{1\%,5\%}$, respectively, compared to the training data. $F(..)$ is the known distribution of p. We measured mean phoneme duration and phoneme duration variance and fitted it to a normal (Gaussian) distribution.

   $$n\_toolong01 = ||\forall_{dur_i \in W_1} dur_i \geq F(\eta_{1\%})|| \tag{7.12}$$

2. *n_toolong01_norm*, *n_toolong05_norm*: *n_toolong01*, *n_toolong05* normalised by speaking rate:

   $$n\_toolong01\_norm = ||\forall_{dur_i \in W_1} \frac{1}{\alpha_c} dur_i \geq F(\eta_{1\%})|| \tag{7.13}$$

3. *n_tooshort01*, *n_tooshort05*, *n_tooshort01_norm*, *n_tooshort05_norm*: See above for too short durations

4. *sequence*: Number of sequences of too long (too short) followed by too short (too long) phones within one utterance.

5. *avg_tempo*: The average speaking rate, estimated only on those utterances that were recognised with high confidence.

6. *stdev_tempo*: The standard deviation of the speaking rate (w.r.t. to average speaking rate of last n utterances).

7. *diff_tempo*: The absolute difference between the average and the current speaking rate.

$$diff\_tempo = \alpha_c - \alpha_c^{avg} \tag{7.14}$$

8. *tempo*: Current speaking rate, see Equation 7.11

To show the relation between the features chosen and the correct/mis-recognised classification some box-and-whiskers plots are shown. Box-and-whiskers plots are a way to look at the overall shape of the data. The central box shows the data between the 'hinges' (roughly quartiles), with the median presented by a line. 'Whiskers' go out to the extremes of the data, and very extreme points are shown by themselves [Ven97]. The plots of the features *n_toolong05_norm*, *n_tooshort05_norm* and *avg_tempo* obtained from the training data of the NN are shown in Figures 7.4, 7.5 and 7.6, respectively.

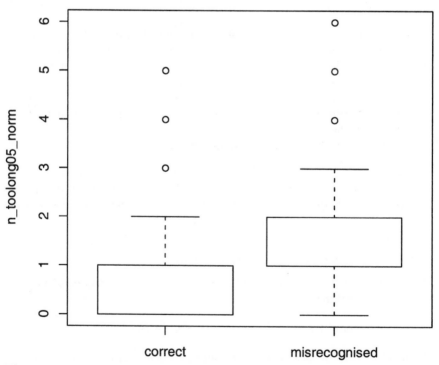

Fig. 7.4. Box-and-whiskers plot for feature *n_too_long05_norm*

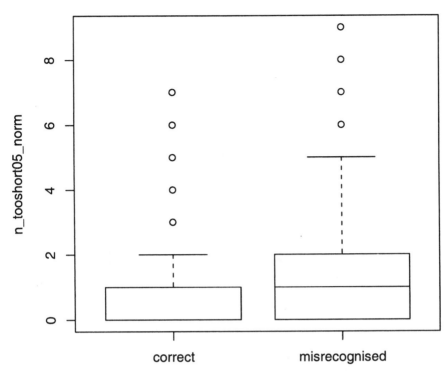

Fig. 7.5. Box-and-whiskers plot for feature $n\_too\_short05\_norm$

Although several outliers are present we can see a good distinction capability between the correct/misrecognised classes of the corresponding feature. In Figures 7.4 and 7.5 the median coincides with the upper bound of the box.

7.4.2 Additional Features

In addition to the duration-based features described above some more features that have proven to be useful for CMs in the past were used. These consist of the following:

1. *feature_direction:* '1' if best word was found in the forward decoding pass, '0' if it was found in the backward decoding pass.
2. *n_frames:* Total number of frames of the utterance (T)
3. *n_phones:* Total number of phones of the utterance (P)
4. *n_frames_nosil:* Total number of frames (without silence) T_{nosil}
5. *first_score:* Acoustic score of best (word) hypothesis W_1:

$$first\_score = p(W_1|\lambda) = \sum_{t=1}^{T} b(o_t)a_{ij} \qquad (7.15)$$

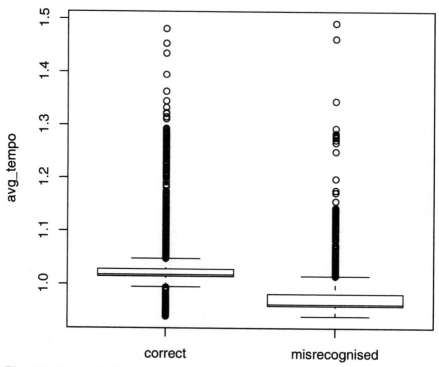

Fig. 7.6. Box-and-whiskers plot for feature *avg_tempo*

6. *first_second*: Difference in acoustic scores between first- and second-best hypothesis:

$$first\_second = p(W_1|\lambda) - p(W_2|\lambda) \qquad (7.16)$$

7. *first_second_l*: *first_second* normalised by the number of frames T
8. *first_second_f*: *first_second* normalised by the score of best hypothesis *first_score*
9. *avg*: Average acoustic score for the N-best hypotheses:

$$avg = \frac{1}{N} \sum_{i=1}^{N} p_i(W_i|\lambda) \qquad (7.17)$$

10. *avg_l*: *avg* normalised by the number of frames T
11. *avg_f*: *avg* normalised by *first_score*
12. *first_avg*: Difference between *first_score* and the average score:

$$first\_avg = p(W_1|\lambda) - avg \qquad (7.18)$$

13. *first_avg_l*: *first_avg* normalised by the number of frames T
14. *first_avg_f*: *first_avg* normalised by *first_score*

15. *first_last*: Difference between first and the last-best hypothesis:

$$first\_last = p(W_1|\lambda) - p(W_N|\lambda) \tag{7.19}$$

16. *first_last_l*: *first_last* normalised by the number of frames T
17. *first_last_f*: *first_last* normalised by *first_score*
18. *first_beambest*: For each frame and all active states the distances between the means of the Gaussian pdfs μ_{s_t} and the feature vectors o_t are computed. The best possible distance, i.e., the minimal one in the beam is compared to the distance found for the state belonging to the optimal state sequence OSS.

$$first\_beambest = \frac{1}{T_{nosil}} \sum_{t=1}^{T_{nosil}} (min_{s \in beam}[\mu_{s_t} - o_t]$$
$$-[\mu_{OSS_t} - o_t]) \tag{7.20}$$

19. *first_beambest_zeros*: The number of frames for which the score difference (see *first_beambest*) is zero:

$$first\_beambest\_zeros = \frac{1}{T_{nosil}} \sum_{t=1}^{T_{nosil}}, \tag{7.21}$$

with $T_{nosil, \in (first\_beambest=0)}$

20. *first_beambest_largest*: The area of the largest continuous difference of *first_beambest* (in terms of subsequent frames)
21. *best*: The best possible score in the beam:

$$best = \sum_{t=1}^{T} min_{s \in beam} b(o_t) \tag{7.22}$$

22. *first_best*: See *first_beambest*, taking into account the transition probabilities:

$$first\_best = \frac{1}{T_{nosil}} \sum_{t=1}^{T_{nosil}} (min_{s \in beam}[\mu_{s_t} - o_t]a_{ij}$$
$$-[\mu_{OSS_t} - o_t]a_{ij}) \tag{7.23}$$

23. *worst_phonescore*: The worst phone score in the best hypothesis, except silence:

$$worst\_phonescore = max_{p \in W_1} p_{phone} \tag{7.24}$$

24. *worst_phonescore_b*: *worst_phonescore* normalised by the best possible phone score in beam p_{best}:

$$p_{best} = \sum_{dur_i} [min_{beam} p_{i,t}] \tag{7.25}$$

$$worst\_phonescore\_b = \frac{1}{p_{best}} worst\_phonescore \tag{7.26}$$

25. *avg_phonescore*: The average phone score in the best hypothesis, except silence:

$$avg\_phonescore = \frac{1}{P} \sum_{i=1}^{P} p_{i,phone} \qquad (7.27)$$

26. *stdev_phnscore*: The change of score within one phone:

$$stdev\_phnscore = \frac{1}{T_{\in phone}} \sqrt{\sum_{t \in phone} (p_{phone_t} - p_{phone_{avg}})^2} \qquad (7.28)$$

27. *worst_phone_best*: The difference between the best possible and worst phoneme score in the best hypothesis:

$$worst\_phone\_best = p_{best} - worst\_phonescore \qquad (7.29)$$

28. *worst_frame_score*: Worst frame score for all frames in the best hypothesis:

$$worst\_frame\_score = max_{t \in W_1} p_{frame} \qquad (7.30)$$

29. *worst_frame_score_b*: *worst_frame_score* normalised by the best possible frame score in beam $p_{t,best}$:

$$p_{t,best} = [min_{beam} p_t] \qquad (7.31)$$

$$worst\_frame\_score\_b = \frac{1}{p_{t,best}} worst\_frame\_score \qquad (7.32)$$

30. *best_in_beam*: The sum of the differences between the best frame scores in the beam and the best hypothesis:

$$best\_in\_beam = \sum [min_{beam} p_{frame_t}] - p_{frame_{OSS}} \qquad (7.33)$$

31. *best_in_beam_l*: *best_in_beam* normalised by the number of frames
32. *avg_framescore*: Average acoustic score for all frames in the best hypothesis, except silence:

$$avg\_framescore = \frac{1}{T_{nosil}} \sum_{i=1}^{T_{nosil}} first\_score \qquad (7.34)$$

7.4.3 Combining the NN Classifier with Speaker Adaptation

In Section 6 we have seen that the weakness of unsupervised adaptation schemes is that they often fail if the baseline performance of the recogniser is too low and too many misrecognitions occur. Since the adaptation has to rely on the recogniser output to be the spoken word, misrecognitions cause the wrong models to be adapted. If this happens repeatedly, the performance may decrease. At this point CMs can be applied to judge how reliable the recogniser result was and accept only those utterances for adaptation that were recognised with high confidence. There has been some interest in this recently. In Feng [Fen95] CMN was performed and combined with a HMM parameter adaptation approach with the CM consisting of the weighted sum of the ML score for each decoded word, a differential ML score (first-best minus second-best) and the duration of the decoded word. Whenever this was above an empirically determined threshold, the utterance was used for adaptation (using Baum-Welch re-estimation). The final parameters were then calculated as the weighted sum of the original parameter and the re-estimated one weighted by the confidence measure and an additional factor to adjust its influence. Anastasakos [Ana98] uses two CMs to guide unsupervised speaker adaptation, one acoustic CM, which is the phone or word posterior probability, and a CM derived from the LM, containing the number of competing hypotheses that end at the same time. Two thresholds were used, one for the LM score and one for the acoustic score. Words whose LM CM was below a certain threshold were rejected, except for those words whose acoustic score was above the acoustic threshold. Similarly those words that were accepted by the LM CM but whose acoustic score was low, were discarded from the adaptation data. In [Mor00] a LR-based CM is used to either accept or reject utterances for adaptation. In [Hom97, Cha01] the difference in likelihood between the first and second-best hypothesis is used to do so. Charlet [Cha01] further investigates a ranking of the words of the hypothesis according to their confidence value and weights them during adaptation accordingly. In her experiments selecting the data for MAP adaptation achieves better results than weighting it. She further states that the influence of the CM on adaptation depends a lot on the adaptation method itself. If a lot of prior information is used during adaptation, the adaptation might not have such a big influence and thus the influence of the CM might also be reduced. In [Pit00] a CM based on refined posterior probabilities that were derived from word graphs to select reliable frames for adaptation was used. Gao and his co-authors [Gao00] use a 'state quality measure' as a CM that basically counts how often a state was assigned to the correct path. This measure is then used to penalise or boost the corresponding segments of speech during adaptation.

Also in the approach presented here, the CM was incorporated into our recogniser so that it guides the adaptation as depicted in Figure 7.7. We call this *semi-supervised adaptation*. The approach is still much more flexible than the classical supervised adaptation schemes, since it still does not have to be known, what was spoken. But we also do not simply use all utterances for adaptation; instead we choose only those that were recognised correctly with high confidence.

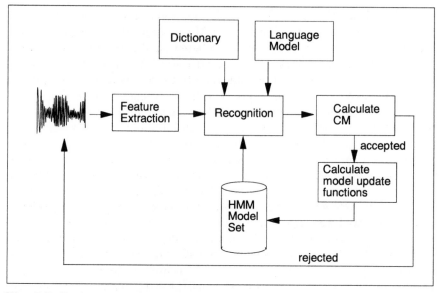

Fig. 7.7. Semi-supervised adaptation procedure

7.5 Experiments and Results

7.5.1 Evaluation of the NN Classifier

First the NN classifier was tested independently from the recogniser, just using the test pattern file. The test pattern file included the confidence features that were extracted from each utterance in the test set, using the standard recogniser. The experimental setup was the one described in Appendix A. The test set consisted of 35 speakers and contained 53.3% of OOV words. The NN had three output nodes. The first two were used to decide whether the word was accepted, the third one was used to identify OOV words.

When using the full set of features, the results that are listed in Table 7.1 were achieved. The first row shows the recognition result if no CM was ap-

Table 7.1. Classification results of a NN using 45 confidence features

	CC	C_a	F_a	C_r	F_r
baseline (SI)	–	35.3%	64.7%	–	–
NN	93.9%	32.9%	3.7%	61.0%	2.5%

plied at all. The very poor baseline performance can be explained by the high number of OOV words in the test set. The second row shows the results after the CM was applied. The NN correctly rejects more than 60% of the utterances (C_r) although at the cost of rejecting 2.5% of the correctly recognised ones (F_r). 3.7% of the utterances were accepted although they were misrecognised (F_a). This means, the F_a and thus misrecognition rate can be reduced by more than 94%. It may not seem realistic to include so many OOV words into the test set, but such a high amount of OOV words was used, to test the NN's capability to identify OOVs. This information is shown in Table 7.2. In 94.9% of the F_a-cases of the baseline system the NN classified OOV words correctly. In dialogue systems it could be beneficial for the course of the dialogue, not just to classify a word or sentence as misrecognised, but also to know if it was an OOV word or not. This knowledge can greatly influence the following dialogue steps.

Table 7.2. OOV classification results for the NN with 45 features

	corr	wrong
3 out	94.9%	5.1%

If we look at the large number of OOV patterns in the test set, we see that 77.3% of the 'misrecognised' patterns were OOV ones. So if we simply classify all misrecognised words as OOV, we would be wrong in 22.7% of the cases on this test set. That means that our NN is much better than simple guessing and therefore than the trivial classifier.

In order to investigate the influence of the duration-based features, separate NNs were trained, one that used only features that were mainly related to the acoustic score (described in Section 7.4.2) and a second one that used the duration-based features only (described in Section 7.4.1).

Two features were used in both sets, the number of frames (*n_frames_nosil*) and the number of phones (*n_phones*). Figure 7.8 shows the results for both, the duration-based and the acoustic score-related features.

The acoustic score-related features ('acoustic score') achieve a CC rate of 91%. The CC rate for the duration-based features ('dur') is slightly worse, 88.5%. The F_a rate is better for the acoustic score features, the F_r rates are better for the duration features. The combination ('all') finally achieves 93.9% CC. This result indicates that there is some information kept in the duration-based features that cannot be covered by the traditional acoustic score-related features. It shows that combining the newly introduced features with the traditional acoustic-score related features can improve the CC results.

Even though the performance of the duration-based features alone is slightly worse than using all or the acoustic-score features, it is still satisfying. Using the duration features only would have the advantage that only the first best hypothesis needs to be computed by the recogniser as opposed to the ten best hypotheses that are required for the acoustic score-related features. This could save some time for decoding, which might be an important factor depending on the application. Furthermore, it is assumed that

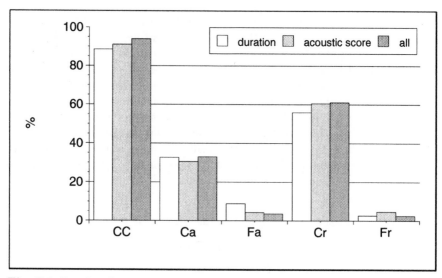

Fig. 7.8. Results of acoustic score and duration-based features

the duration-based features are more robust to changes in the recogniser settings, for example in the front-end. While the front-end surely has an effect on the acoustic scores, its influence on the phoneme duration rather is judged modest. However, this is just an assumption and we were not able to prove it due to the lack of other data.

7.5.2 Semi-supervised Adaptation

The NN was integrated into our recognition system using weighted MLLR adaptation. Again the command corpus was used for testing.

More and more OOV words were added step-wise to the testing data to investigate their influence on unsupervised adaptation. Utterances from the address corpus were recognised together with the command corpus, so that we knew that they were going to be misrecognised, because they were not included in the recogniser's vocabulary. Figure 7.9 shows the results after adding the addresses a different number of times (corresponding to 80, 200 and 400 OOV utterances per speaker. Remember that the command corpus comprised approximately 200 utterances per speaker). The baseline result is depicted in the first bar ('base'). The second bar shows unsupervised adaptation ('unsup') and the third bar shows semi-supervised adaptation ('semi-sup'). These blocks are repeated for different numbers of added OOV words. It can be seen that the semi-supervised approach is always better than the unsupervised approach. The more OOV words are included, the better is the semi-supervised performance compared to the unsupervised approach, because more and more OOV words are used for adaptation in the unsupervised case. However the unsupervised approach is better than the baseline for up to 80 OOVs, which indicates its robustness even for a large number of OOV words. A further advantage of the semi-supervised approach can be seen if we look at Table 7.3, which shows the number of model updates done by adaptation conducted for the experiments shown in Figure 7.9.

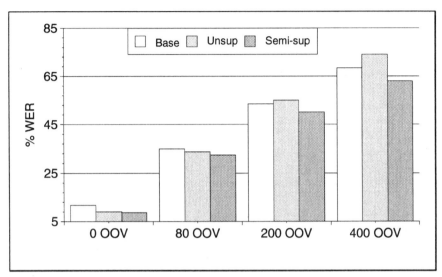

Fig. 7.9. Unsupervised vs. semi-supervised adaptation for different number of OOV utterances

Table 7.3. Number of HMM model updates done by adaptation

	0 OOV	80 OOV	200 OOV	400 OOV
unsupervised	3162	4494	6492	9822
semi-supervised	1354	1338	1265	1457

It can be seen that for the semi-supervised approach the number of HMM model updates done by adaptation remains more or less constant, even if the number of OOVs and thus the total number of words to be recognised increases. Using the unsupervised approach that accepts any utterance for adaptation, the number of updates increases a lot without yielding better recognition rates. Since we should save computation time whenever it is possible, the use of the semi-supervised approach is beneficial, since adaptation is only conducted when 'required'.

7.6 Summary

In this chapter, the CM that was adopted in this research work was presented. A two-step procedure that collects a set of features at recognition time and uses a NN classifier to process the features to make a decision whether to accept or reject the utterance was used. Some new features that are mainly related to phoneme duration statistics were explained and combined with traditional features that were mostly related to the acoustic score. The NN using the full set of features achieved a classification rate of 93.9%. Simultaneously we succeeded in identifying OOV words in 94.9% of the F_a-cases of the baseline system. By training separate NNs for the acoustic score-related and duration-based features, it could be shown that some information that is not kept in the acoustic features can be captured by the duration features. It was thus possible to improve the CC rate of the classifier by introducing the new features. Even though the performance of the duration-based features alone was slightly worse than the complete set of features, for certain applications such a NN is preferable because it is computationally less expensive.

The NN classifier was integrated into our adaptation scheme, yielding a semi-supervised adaptation. The results show that the use of the CM does support adaptation, especially if many misrecognitions occur, which was simulated by adding a large number of OOV words to the test set. Furthermore it was shown that using the semi-supervised approach kept the number of HMM model updates done by the adaptation at a constant level even if a large number of OOV utterances were added to the test without any loss in performance. For large amounts of OOV words performance was better than for the unsupervised approach. Since this can also save a lot of computation time it is of great interest for many applications.

8 Pronunciation Adaptation

The techniques of ASR and ASU have reached a level where they are ready to be used in products. Nowadays people can travel easily to almost every place in the world and due to the growing globalisation of business they even have to. As a consequence public information systems are frequently used by foreigners or people who are new to a city or region. If the systems are equipped with a speech interface, supporting several languages is extremely costly since for each language AMs, pronunciation lexica and LMs need to be built, all of which require large amounts of data for each language. The number of languages has to be restricted somehow and as a result many people will choose one of the supported languages and speak with a foreign accent. Also the number of growing speech enabled Internet applications or various content selection tasks might require the speaker to say titles or the like in a foreign language. Most of the systems nowadays are not able to handle accented speech adequately and perform rather badly, compared to the performance achieved when recognising speech of native speakers, see [Wit99a, Cha97, Bon98].

While it was possible to capture some of the pronunciation variation in high complexity HMM models for native speakers, this will not be possible to the required extent for non-native speech, as was already indicated by the findings of Jurafsky [Jur01]. Also acoustic adaptation alone as described in Chapter 6, will not be sufficient to deal with accents. This is also supported by Woodland [Woo99].

State-of-the-art speech recognition systems use a pronunciation dictionary to map the orthography of each word to its pronunciation, as we have seen in Chapter 5. Usually, a 'standard' pronunciation (also called *canonical pronunciation* or *base form*), as it can be found in published pronunciation dictionaries, is used. These canonical pronunciations show the phonemic representation of a word; that is, how it should be pronounced if it is spoken in isolation. This is also called *citation form*, see [Pau98]. In isolated speech this canonical pronunciation might come close to what speakers actually say, but for some words there is a mismatch between this standard pronunciation and its phonetic realisation. For continuous or even spontaneous speech, the problem gets more severe. Due to co-articulation effects the pronounced words tend to deviate more and more from the canonical pronunciation, especially at higher speaking rates. A transcription of spontaneous American English speech (Switchboard) revealed 80 variants of the word 'the', see [Gre99] that are certainly not captured in any standard American English pronunciation dictionary. Jost and his co-authors [Jos97] estimated that in spontaneous speech 40% of all words are not pronounced as in the standard pronunciation dictionary. Also the pronunciation is very much speaker-specific, so that it is not possible to determine one correct pronunciation. Many approaches were presented that try to account for this variability by including more than one pronunciation in the dictionary. Most of the research done in pronunciation modelling, however, has focused solely on the derivation of native variants.

Since the problem is much more severe in the case of non-native speakers speaking with a more or less strong accent, this particular problem will be considered in the following sections. For this purpose a non-native database is closely examined and the effect of including specialised non-native pronunciations in the dictionary on recognition accuracy is demonstrated. So far the biggest problem when modelling non-native pronunciation has been to obtain accented speech data bases. In this chapter a very flexible method is proposed that makes it possible to derive pronunciation rules for non-native speakers without using any accented speech data at all. It needs native data of the two considered languages only and is thus applicable to any combination of languages. Before these methods are described in detail, an overview of the state of the art in pronunciation modelling is given in Section 8.1.

8.1 The State of the Art in Pronunciation Modelling

The dictionary relates the orthography of the words that are known to the system – representing the desired recognition output – to their pronunciations – representing the input by the speaker. Of course it would be desirable to automatically derive the pronunciation of a word from its orthography without the need of human interference. The manual creation and/or correction of a pronunciation or segmentation of an utterance requires expert knowledge and is a very time consuming and costly task especially for large vocabularies of several 10,000 words. Adding new words to the vocabulary is problematic, since the expert needs to be consulted again. A further problem is that different trained specialists often do not produce exactly the same transcription for the same utterance or consider the same variants for the same word as important.

There is a consensus that including multiple pronunciation variants is important for ASR systems. Lamel and her co-authors [Lam96] showed that a careful design of the pronunciation dictionary influences recognition performance. In [AD99] a further investigation showed that not including certain variants introduces recognition errors. The main problem is that the more alternatives that are incorporated in the dictionary, the higher is the probability that one of them comes quite close to the pronunciation of a different word. Thus the confusability in the dictionary increases. In many publications the positive effect of including variants was more than nullified by the increased confusability, e.g. in [Noc98, MB98, AD99]. Lamel and Adda-Decker [Lam96, AD99] furthermore found a (language dependent) relation between the word frequency and the number of pronunciation variants used, such that the number of variants increases with the frequency of the words.

Also McAllaster [McA98] showed that, if all pronunciations in the corpus are in the dictionary, a dramatic decrease in error rate can be achieved. He showed this using simulated data that were automatically generated using the pronunciation dictionary. Even if pronunciations were added that were shared by different words, the error rates could drastically be reduced for a spontaneous speech task. But he additionally found that if more variants than actually occurred in the test data are included in the dictionary, performance declined. So the consequence is that the dictionary needs to accurately reflect the range of phonetic variation observed, otherwise performance is impaired. If the same test was conducted on real data, an increase in error rate was observed always, even if only variants that were known to appear in the test data were added. He sees the main reason in the fact that the HMM models used were trained on data aligned with the canonical pronunciations, thus resulting in 'diffuse' models. So both, [McA98] and [AD99] argue that introducing variants for the alignment of data that are used for HMM training is necessary because it will produce more accurate AMs. But it needs to be mentioned that there is considerable controversy over whether training with more accurate transcriptions is better than training with the canonical transcription only.

The majority of current speech recognisers use HMMs. High complexity models, such as triphones with many mixtures, are already capable of modelling pronunciation variation and co-articulation effects to a certain extent, see [Hol99, AD99, Ril99]. Jurafsky and his co-authors investigated this issue more closely [Jur01] and found that on one hand, some of the variation, such as vowel reduction and phoneme substitution can indeed be handled by triphones, provided that more training data for the cases under consideration are available for triphone training. On the other hand, there are variations like syllable deletions that cannot be captured by increased training data. Recently He and Zhao [He01] found that for non-native English speakers triphones perform worse than monophones. Ravishankar [Rav97] also states that, inferior pronunciations do not always cause misrecognitions because they can to a certain extent be handled by the AMs, but they will lower the acoustic likelihood of the sentence in which such a word occurs and will increase the chances of an error elsewhere in the utterance. This is also confirmed by Fosler-Lussier [FL99], who found in experiments that not all words for which the actual pronunciation differs from the one in the dictionary are misrecognised. He states that not all variation can be appropriately captured this way, especially if spontaneous speech is considered. Here factors such as word frequency and speaking rate play an important role. If the speaking rate is very high and the words are very frequent the syllabic distance from the canonical pronunciation increases for many syllables. From the information-theoretic viewpoint this makes sense, since high-frequency words are more predictable. Thus more variation is allowed in their production at various speaking rates because the listener will be able to reconstruct what was said from the context and a few acoustic cues.

Greenberg [Gre99] investigated the influence of speaking rate on pronunciations and found that at high speaking rates, more deletions take place. Also [FL99] confirmed that infrequently used words that are of higher information valence tend to be pronounced canonically while frequently used words (like pronouns, function word or articles) deviate from the canonical pronunciation quite regularly.

All the results described indicate that variation in pronunciation cannot be handled appropriately solely on the AM level (by increasing the data used for triphone training or using speaker adaptation techniques) but that a modification of the pronunciation dictionary becomes necessary. One of the few publications known to the author that combines acoustic adaptation (using MLLR) with the adaptation of the dictionary for non-native speech is that of Huang [Hua00]. He also assumes that acoustic deviation is an independent but complementary phenomenon and thus independently uses MLLR adaptation for adapting the AMs and enhances the dictionary with accented variants. He shows that the combination of the two methods yields better results than either method alone. His work will be described in more detail in this chapter.

The existing approaches for modelling pronunciation variation can be classified into three broad classes, rule-based, data-driven and combined approaches. They will be presented in the following sections.

8.1.1 Rule-Based Approaches

In rule-based approaches a set of pronunciation rules is used to transform the standard pronunciation into a pronunciation variant. They can for example be of the form

$$/l \ @ \ n/ \rightarrow /l \ n/$$

to account for the @-deletion at word endings in German. The rules are derived using linguistic and phonetic knowledge on what kind of pronunciation variation occurs in the kind of speech considered. These rules are then applied to the baseline dictionary. Thus for the entries in the dictionary a set of alternative pronunciations is obtained and added. The advantage of this approach is that it is completely task-independent, since it uses general linguistic and phonetic rules and can thus be used across corpora and especially for new words that are introduced to the system. The drawback however, is that the rules are often very general and thus too many variants are generated, some of which might not be observed very often. As we have already seen, too many variants increase confusability. The difficulty is then to find those rules that are really relevant for the task.

An example of rule-based approaches is that of Lethinen [Let98], who showed that using very rudimentary rules, which were obtained by segmenting a graphemic string and simply converting the segments according to a grapheme-to-phoneme alphabet for German and using all transcriptions generated together with their application likelihoods very often receives a higher ranking than the canonical transcription. Also Wiseman and Downey [Wis98, Dow98] show that some rules have effects on the recognition accuracy while others don't.

In several publications, [Wes96a, Wes96b, Kip96, Kip97] present their work on the Munich AUtomatic Segmentation system (MAUS). Pronunciation variants that were needed for segmenting the training data are generated using a set of rules. In [Kip97] this rule-based approach is compared to a statistical pronunciation model that uses micro pronunciation variants that apply to a small number of phonemes. These rules are determined from a hand-labelled corpus. The latter model achieves higher agreement with manual transcriptions than the rule-based approach.

8.1.2 Data-Driven Approaches

In data-driven approaches, the alternative pronunciations are learned from the speech data directly, so that it is possible to also compute application likelihoods from this data. These likelihoods are a measure for how frequently a certain pronunciation is used. Additionally, in this case only pronunciations are generated that are really used. However, this is very much corpus dependent and variants that are frequently used in one corpus do not necessarily have to be used in another corpus as well. Another problem is that this approach lacks any generalisation capability.

A method that is often used to derive the variants directly from the speech data is to use a phoneme recogniser. A brief description of a phoneme recogniser is given in Section 8.1.2. The problem with this approach is to cope with the large number of phoneme errors that are introduced by the recogniser. The highest achievable phoneme recognition rates without using any further constraints for the recogniser have been between 50 and 70% in the past [dM98]. Further restrictions could be made by using phoneme bi- or trigrams, but since the goal is to find unknown variants, the freedom of the recogniser should be as high as possible.

Hanna [Han99a] addressed the problem of phoneme recognition errors by assigning different probabilities to insertions, deletions and substitutions. This was done to avoid equally probable transcriptions if all insertion, deletion and substitution probabilities are assigned the same values. To compute separate probabilities, an iterative DP scheme and a confusion matrix were used. While the most significant substitutions were retained, the number of insignificant ones was reduced. In [Han99b] additionally pronunciations resulting from co-articulation were removed first.

Wester [Wes00a], uses decision trees to prune the variants generated by a phoneme recogniser. Amdal [Ama00], uses a measure they call association strength between phones. They use statistics on co-occurrences of phones and use these for the alignment of the reference and alternative transcriptions, which was generated using a phoneme recogniser. They create rules for speakers separately and then merge them into one dictionary. Although this is a promising approach for discarding phoneme recognition errors, it requires a lot of data. Williams [Wil98c], uses CMs to select reliable pronunciations and Mokbel [Mok98] groups variants that were obtained by a phoneme recogniser and represents each group by one transcription.

Phoneme Recogniser. As described above, often a phoneme recogniser is used to derive pronunciation variants from speech data directly, if the only available information source is the speech signal and the orthography of the word that was spoken. The phoneme recognition result then provides a possible pronunciation for this word. A phoneme recogniser is a special case in recognition. The dictionary does not consist of words as usual, but of phonemes only. Therefore no words but only phoneme sequences can be recognised (that can optionally later be mapped to words). An example of an English phoneme recogniser is depicted in Figure 8.1. The search is either not restricted at all, so that arbitrary phoneme sequences can be recognised or it is restricted by a phoneme bi- or tri-gram LM. Then certain phoneme sequences are favoured by the LM. Although providing the phoneme sequence best fitting the speech signal, and thus often achieving higher phoneme scores than in word recognition tasks, phoneme recognition rates are usually not very high, between 50-70%. But human expert spectrogram readers were also only able to achieve a phoneme recognition rate of around 69% [dM98]. This again shows that the context and meaning of a word or sentence plays an important role for the recognition of speech both for humans and machines. A phoneme recogniser was used for the derivation of non-native pronunciation variants and the whole generation procedure will be described in Section 8.4.

8.1.3 Combined Approaches

It seems to be a good solution to combine the rule-based and data-driven approaches that is to use speech corpora to derive a set of rules. Therefore phenomena really occurring in the speech can be covered but still the possibility to generalise to other tasks and corpora can be retained. Of course, the rules derived still depend on the corpus used. An example for such a combined approach is given by Cremelie and Martens [Cre97, Cre99]. A simple left-to-right pronunciation model was built by a forced alignment using the

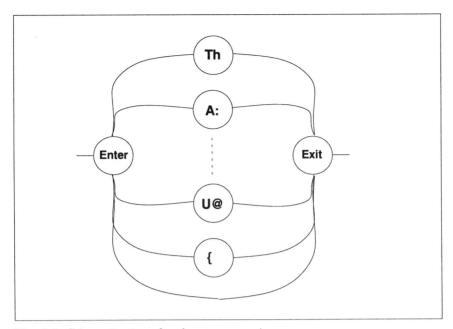

Fig. 8.1. Schematic view of a phoneme recogniser

standard transcription only. Then at all possible positions, deletion, insertion and substitution transitions were inserted and a second alignment using this model was conducted. From this, a number of candidate rules were generated by finding the differences between the reference and this new transcription. Then the rules were pruned and only the most likely ones were kept. So pronunciation variants are obtained together with their application likelihoods. A detailed analysis of the results revealed that most of the improvement was caused by the co-articulation rules. This was also shown in [Yan00a, Yan00b].

Another way to capture linguistic knowledge is to grow decision trees using a set of linguistic and phonetic questions, see [Ril96, Ril99]. Similar to the method proposed here, [Ril96, Ril99, Hum96] used decision trees to learn a phone-to-phone mapping from the canonical pronunciations to the one obtained from a database. While [Ril96, Ril99] extracted the variants from hand-labelled corpora, [Hum96] obtained the variants by re-transcribing the speech data with SI models and deriving typical substitutions. However, they restricted the possible substitutions to vowels only. Always, the decision trees were then used to generate the alternative pronunciations for the baseline dictionary.

In general it can be stated that within-word variation which frequently occurs in CSR is not so easy to derive just from a dictionary because it depends very much on the speaking style, speaking rate and the speaker. Cross-word variations, as for example in 'going to', which is often pronounced like 'gonna', are easier to derive, because the changes that are made to the pronunciation of one word depend on the preceding and following word to a high extent (although still depending on the speaker as well). In contrast to [Cre99, Yan00a], [Kes99] found that modelling cross-word variation actually increases the WER if taken in isolation, whereas together with within-word variation it improves the performance.

8.1.4 Miscellaneous Approaches

Apart from modelling pronunciation variation on the dictionary level it can also be modelled on the HMM level. Some approaches exist that try to account for the variability in pronunciation by the development of specialised HMMs that are trained or tailored for fast, slow or emotional speech, [Pfa98, Fal99, Pol98, Eid99, Hei98, Wom97]. Another widely used approach is to use NNs to model pronunciations, [MB98, Fuk98, Fuk97, Des96].

A common way to deal with cross-word variation is to use *multi-words*, which are joined together words that are added as entities to the dictionary. Those word sequences that occur frequently in spontaneous speech should be covered without increasing the confusability with other words too much. When added to the LM, they are often assigned the same probability as the single word sequences. Multi-words were used e.g. by [Slo96], who generated variants for words and multi-words by running a phoneme recogniser that was constrained by a phoneme bi-gram. Relevant variants were determined by a CM. The sets of multi-words were determined based on the frequency of occurrence in the corpus. Both adding variants and multi-words could decrease the WERs. This multi-word approach was also used by [Rav97, Fin97, Noc98].

8.1.5 Re-training the Acoustic Models

Most of the work so far presented, concentrated on using pronunciation variants during recognition only. Even though there is a common agreement that the modelling of pronunciation variation is an important topic, the improvements in WER are rather modest in most of the cases. In [Wes00b] a detailed analysis of the results was done, revealing that there are many differences in the recognition result that are not reflected by the WER.

Another reason might be that usually the training of the acoustic models is done using only the canonical pronunciation for the alignment of the training data. The reason why it seems reasonable to re-train the acoustic models is that if better transcriptions containing the actually spoken variants are used, the number of alignment errors can be reduced and this in turn will result in sharper models. An example of this was given already in Section 5.2. In many publications, e.g. [Byr97, Slo96, Sar99, Fin97, Kes99] it has proven to be beneficial to do a re-training of the acoustic models, using re-transcribed data. An exception to this is [Hol99], where a re-training of the acoustic models did not help.

When pronunciation variations are considered during recognition another problem arises. If the LM remains unchanged (that means only the orthography of a word is used in the LM), but for decoding a dictionary containing all pronunciation variants is used, the consequence would be that all variants are assigned the same a priori probability. This does not reflect reality. So it would be better to include the variants with their corresponding probabilities directly into the LM. However this increases complexity and requires a corpus labelled with variants to obtain the LM probabilities for each of the variants. Pousse [Pou97], showed that using contextual variants in a bi-gram LM outperforms a LM that includes the orthographic representation of the words only, but of course the complexity of the LM increased a lot. In [Kes99], it was also shown that incorporating the variants into the LM is beneficial.

To summarise the state of the art in pronunciation modelling, it can be stated that it is considered to be an important topic. However, results are not as good as expected in most of the cases. One reason is the potentially increased confusability, another one the use of 'diffuse' HMM models during decoding. Some techniques, while improving the results on one database, do not succeed on other databases, so it seems that considerable research is still necessary concerning this topic.

8.2 Pronunciation Modelling of Accented and Dialect Speech

The main focus of the research in pronunciation modelling has so far been native speech. When non-native or accented speech is considered, some more problems arise. In the following description the native language of a speaker will be called *source language* and the language that he is trying to speak will be called *target language*. While in native speech, basically only insertions, deletions and substitutions within the phoneme set of that particular language have to be considered, for non-native speech phonemes of the source languages are also used by many speakers. Witt [Wit99b], states that in general learners of a second or third language tend to apply articulatory habits and phonological knowledge of their native language. This is one reason why many approaches assume a known mother tongue of the speaker, because certain characteristics of the accented speech can then be predicted easily, see [Bon98, Fis98, Fra99, Tra99, Her99, Wit99a, Tom00]. If no similar sound of the target language exists in the source language, speakers tend to either insert or delete vowels or consonants, in order to reproduce a syllable structure comparable to their native language. Furthermore, non-native speech is often characterised by lower speech rates and different temporal characteristics of the sounds, especially if the second language is not spoken fluently.

However, considering the intelligibility of non-native speech spectral characteristics, such as F_2 and F_3 frequency locations, [Wit99a] found that the latter have a stronger influence than the temporal ones. These frequency locations are influenced by changes in the tongue restriction centre (narrowing the vocal tract due to the tongue). In contrast to this, F_1 does not seem to have such a strong influence on the intelligibility. F_1 changes with the vocal tract. So it seems that the tricky part when speaking in a foreign language are the tongue movements.

As mentioned above, Humphries and his co-authors [Hum96] explicitly modelled accented variants and achieved major reductions in WER.

Huang and his co-authors [Hua00] are one of the few researchers who investigated the combination of MLLR speaker adaptation and the modification of the pronunciation dictionary. They exploit the fact that between speaker groups of different accents some clear tendencies of e.g. phoneme substitutions can be observed. They obtained syllable level transcriptions using a syllable recogniser for Mandarin and aligned these with the reference transcriptions to identify error pairs (mainly substitutions were considered). These were used to derive transformation rules. New pronunciations were generated using these transformation rules and added to the canonical dictionary. Using the extended dictionary and MLLR speaker adaptation alone, improves the results. When both methods are combined even better improvements are achieved, which indicates that acoustic deviation from the SI models and pronunciation variation are at least partly independent phenomena.

This was also shown in an approach that combined pronunciation variant modelling with VTLN, see [Pfa97]. Pfau and his co-authors use HMM models specialised for different speaking rates and tested the effect of VTLN and pronunciation modelling on these models. While both VTLN and the use of pronunciation variants in the dictionary could improve the results, the best results were achieved when both methods were combined.

8.3 Recognising Non-native Speech

In this section, some experiments are presented that tested the effect of adding relevant non-native pronunciation variants to the pronunciation dictionary for a non-native speech database. Recognition rates were computed on the Interactive Spoken Language Education corpus (ISLE) (see [ISL]). The speakers contained in this database are German and Italian learners of English. The database was recorded for a project that aimed at developing a language learning tool that helps learners of a second language to improve their pronunciation. The ISLE corpus is one of the few corpora that exclusively contains non-native speakers with the same mother tongue, in this case German and Italian. Often, if non-native speech is contained at all in a database, it covers a diversity of source languages but only very few speakers with the same source language. A more detailed description of the ISLE corpus can be found in Appendix A.

However, the ISLE database is not optimal in the sense that very special sentences were recorded. As already mentioned, the goal of the project was to develop a language learning tool with special focus on evaluating the pronunciation quality. So the sentences were carefully chosen so as to contain as many as possible phoneme sequences that are known to be problematic for Germans and Italians. The database is divided into blocks and while some of the blocks contain very long sentences that were read from a book, other parts contain only very short phrases, like e.g. 'a thumb'. This was especially a problem for the training of the LM for this task. The only textual data that were available were the transcriptions that came with the ISLE database. A bi-gram LM was trained using this data. The LM perplexity was 5.9. We expect sub-optimal results from such a LM as far as the recognition rates were concerned due to the very different structure of training sentences. Furthermore the fact that in contrast to our previous experiments this is a CSR task, required the use of a different recogniser[1], since the previously used one can only handle isolated words and short phrases. The pre-processing of the speech was the same as in all previous experiments. The baseline WER using British English monophone models, trained on the British English Wall Street Journal (WSJ), averaged over twelve test speakers was 40.7%. Unfortunately no native speakers were available to evaluate to what extent the low baseline recognition rates are caused by the foreign accent.

Part of the database is manually transcribed and contains the pronunciations that were actually used by the speakers. These manual labels did not consider the possibility that German or Italian phonemes were used; they used the British English phoneme set only. All these non-native pronunciations were added to the baseline dictionary and tested. The results can be seen in Figures 8.2 and 8.3 for Italian and German speakers, respectively. The number of pronunciations per speaker was 1.2 and 1.8 if the German and Italian rules were applied, respectively. In the baseline dictionary which included a few native variants already, the average number was 1.03. In the following figures, the speaker IDs are used together with the indices '_I' (for Italian) and '_G' (for German). First of all, it should be noted, that the baseline recognition rates for Italian speakers are much worse than for German speakers. This allows one to draw the conclusion that in general the English of the Italian speakers is worse than that of the German speakers in terms of pronunciation errors. This was confirmed by the manual rating that was available for the database and counted the phoneme error rates of the speakers.

It can be seen that the recognition rates can be improved for almost all speakers in the test set if the dictionary is enhanced with the corresponding variants of the respective language (indicated by 'GerVars' and 'ItaVars',

---

[1] HTK3.0 was used for the ISLE experiments, see [HTK]

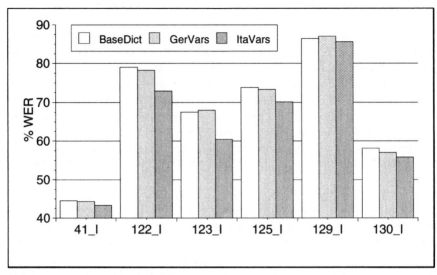

Fig. 8.2. WERs for Italian speakers, using manually derived variants

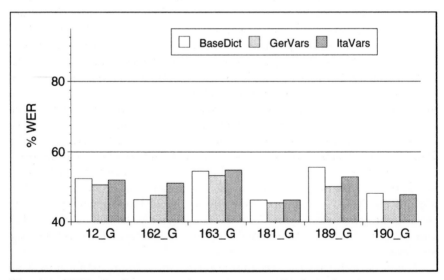

Fig. 8.3. WERs for German speakers, using manually derived variants

respectively). Interestingly, for one German speaker (189_G) the Italian variants also improved the results and for some Italian speakers (41_I,122_I, 125_I, 130_I) the application of the German rules yielded improvements, indicating that the necessary variants depend not only on the mother tongue of a speaker but also on the speaker himself. Please note that manual phonetic transcriptions were available only for two thirds of the data. That means that only for words that occurred in these parts, pronunciation variants were added to the dictionary. For all other words, no variants were included.

In a second step the canonical and the manual labels were used to manually derive a set of pronunciation rules for each source language, thus obtaining a 'German' and an 'Italian' rule set.

For German, the most important rule from the phonological point of view is the inability of Germans to reduce vowels. In English in unstressed syllables, vowels are very often reduced. The vowel that is mostly affected is /@/. In German unstressed vowels are, with few exceptions, pronounced with their full quality. To give an example, the correct British pronunciation of the English word 'desert' is /d e z @ t/. Many German speakers however, pronounce it /d e z 3: t/. The occurrence, i.e., how often the phoneme obeying the rule was involved in errors, was 21.7%.

For Italian, the most important rule is to append a /@/ at word final consonants, thus creating an open syllable, e.g. in the word 'between' /b @ t w i: n @/ for /b @ t w i: n/. The occurrence for this was 13.5%. A more detailed description of all pronunciation rules can be found in Section C.3 in Appendix C and in [Gor01b, Sah01]. The derived rules were used to automatically generate German- and Italian-accented variants from the canonical pronunciations, this time for all words in the dictionary.

First, simply all rules were used to derive the variants, yielding 3.2 variants per word. The corresponding recognition results (in % WER) are shown in the second bar ('AllRules') of Figures 8.4 and 8.5, for the Italian and German speakers, respectively. It can be seen that for four out of five Italian speakers, but only for one German speaker, the baseline results can be improved. The third bar ('ItaRules') shows the results if the Italian rules were used, resulting in 2.5 variants per word. The baseline results can be improved for all Italian speakers and the improvements are even bigger than before. For all German speakers when using these Italian rules, performance decreases drastically. This is exactly as expected, since we consider the Italian rules as irrelevant for the German speakers. When applying the German rules, which are depicted in the fourth bar ('GerRules') and yield 2.0 variants per word, for the German speakers, the results are better than using all rules, but still only for one speaker, the baseline recognition rate can be improved. Interestingly, for some Italian speakers, the German rules can also improve the baseline results, of course not as much as the Italian rules. Finally, when selecting the rules speaker-wise (that means selecting from all German and all Italian rules those that can improve the results for the respective speaker if tested separately) to create the speaker-optimised dictionaries ('optRules'), the biggest improvements can be achieved for all German and four Italian speakers.

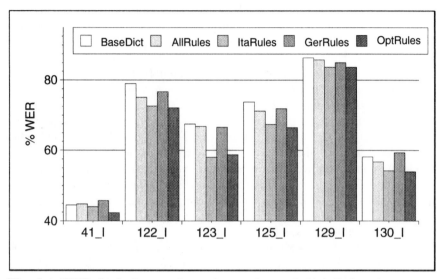

Fig. 8.4. WERs for Italian speakers using speaker-optimised dictionaries

When we compare the results for the Italian and German rule set to the results when the manually found variants were added directly to the dictionary, we can observe that for many speakers the rules perform better than the manual variants. One reason probably is that using the rules, we generate non-native variants for all words in the dictionary, whereas the manual variants were only available for parts of the corpus. This allows one to draw the conclusion that the rule set is able to generate relevant non-native pronunciation variants.

The analysis of the recognition rates shows that a speaker-wise selection of the rules is superior to adding all rules or rules that are typical for a speaker of that mother tongue. This is shown by the fact that some of the German rules improve the results for some Italian speakers and vice versa. We thus conclude that using rules that do not reflect the speaker's articulatory habits can indeed lower the recognition rates, due to the increased confusability. We further conclude that the closer a speaker's pronunciation comes to the native one, the more carefully the rules need to be selected. For most Italian speakers applying all the rules or the Italian rules was already sufficient to improve the baseline. This was not the case for the German speakers, where only the speaker-optimised dictionaries achieved improvements.

As outlined above we assume that on one hand we have to account for the 'non-native phoneme sequences' that might occur, but on the other hand also

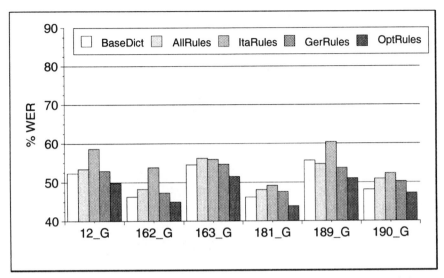

Fig. 8.5. Recognition results for German speakers using speaker-optimised dictionaries

for the model mismatch that will occur for non-native speakers. Even if the same phonemes exist in two languages, they will be different in sound colour. This phenomenon cannot be captured by pronunciation adaptation alone but by acoustic speaker adaptation. So we expect a combination of both methods to be beneficial.

Figure 8.6 shows again the results averaged over all speakers and additionally the cases where the adaptation was applied to the baseline ('BaseDict+MLLR') as well as to the speaker-optimised dictionaries ('optRules+-MLLR'). MLLR can improve the baseline and the manually derived variants ('manVars') but even more when the speaker-specific dictionaries are used. Please note that the results for the combination of MLLR and the optimised dictionaries are bigger than for either method alone; that implies that the improvements are at least partly additive and thus the both methods should be combined.

Even though the results clearly showed the necessity of explicitly modelling non-native pronunciation variation in the dictionary, the rule-based approach we used and which required the manual derivation of the rules from an accented speech corpus, is not appropriate if we want to deal with a variety of source language accents for different languages. Therefore, a new method for generating pronunciation rules was developed and is presented in the next section.

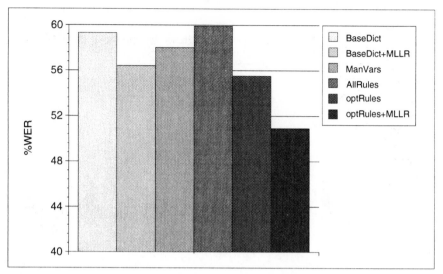

Fig. 8.6. Overall recognition results if speaker-optimised dictionaries and MLLR adaptation were used

8.4 Generating Non-native Pronunciation Variants

Neither the traditional rule-based nor the traditional data-driven methods are suited for the problems we want to tackle, since they are too costly and too inflexible. If we chose the rule-based approach we have to have experts for the various source languages we want to consider, who could generate rule-sets for these source languages when the target language is to be spoken. If we chose the data-driven approach we would have to collect large databases in which several speakers from each source language are recorded speaking the target language. Neither is feasible. This motivated the development of a new approach that has been described in [Gor01a].

The following considerations apply to speakers who are not able to speak the target language at all. They hear the words spoken by native speakers of the target language several times and then try to reproduce them. The basic assumption is that they will try to speak the words the way they hear them. We again use the language pair German-English, but this time the setting is just the other way round. Now the source language is English and the target language is German. This switch was necessary since we wanted to exploit the many repetitions of the same word that are only available in the German command corpus, but not in the ISLE corpus.

Concretely that means that an English speaker will listen to one or several German speakers speaking the same German word, and will then try to repeat it with his English phoneme inventory. This process is shown in Figure 8.7. In this example the English speaker tries to repeat the German word 'Alarm', after having heard it for several times spoken by different German speakers.

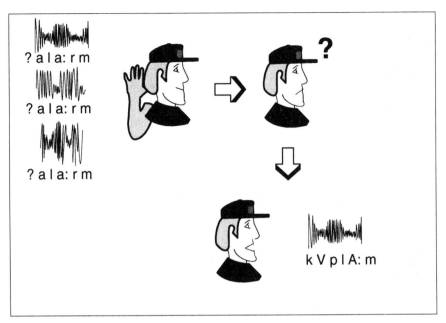

Fig. 8.7. English speaker reproducing German speech

This procedure is simulated by using the HMM models that were trained on the British English WSJ corpus to recognise German speech. This is done using a phoneme recogniser and a phoneme bi-gram LM that was trained on the English phoneme transcription files, so as to reflect the English phonotactic constraints. Using these constraints insertions and deletions that are typical for English are expected to occur.

Recognising the German utterances with the English phoneme recogniser provides us with several 'English transcriptions' for those words. These are used to train a decision tree (as will be explained in Section 8.4.1) that is then used to predict English-accented variants from the German canonical one. This two step procedure is again depicted in Figure 8.8.

The great advantage of the proposed method is that it needs only native speech data, in the case considered native English and native German data. Usually one of the biggest problems is to acquire sufficient amounts of accented data.

Some example phoneme recognition results for the German word 'Alarm' are given in Figure 8.9. In the database the word 'Alarm' was spoken 186 times by 186 different speakers. The left hand side shows the (canonical) German pronunciation (which might be not exactly what was spoken, because the database was aligned with the canonical pronunciation only), on the right hand side some English phoneme recognition results are listed.

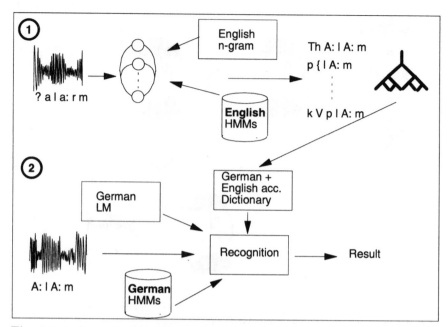

Fig. 8.8. 1: Generating English pronunciations for German words and training the decision tree 2: Applying the decision tree to the German baseline dictionary

Word 'Alarm' spoken by different native speakers of German:	English phoneme recognition results:
? a l a: r m	T h A: l **A: m**
? a l a: r m	sil b al l **A: m**
? a l a: r m	V l aU @ m
? a l a: r m	p { l **A: m**
? a l a: r m	al l aU h V m u:
? a l a: r m	h Q l **A: m**
? a l a: r m	U k b { l **A: u:**

Fig. 8.9. English phoneme recognition result for the German word 'Alarm'

We can see that the second part of the word is often recognised as /l A: m/. For the beginning of the word there seems to be no consistency at all. The reason in this particular case might be fact that the words start with a glottal stop /?/ in German. Nothing comparable does exist in English that's why it is hard to find a 'replacement'. An auditory inspection of the results showed that some English phoneme recognition results, although at first sight seeming to be nonsense, were quite reasonable transcriptions when listening to the phoneme segments. However, there were also many genuinely useless results. By inducing decision trees, the hope is that the consistency that is in our example found for the end of the word can be kept there. For the word beginning however no phoneme sequence occurred twice in the example, which lets us assume that it will be hard for the decision tree to predict any reliable pronunciation at all. For these cases we added several instances of the correct German pronunciation, so that if no English equivalent phoneme sequence can be found, at least the German one will be retained for that part. The best results for the non-native speakers were achieved when two German pronunciations were added to (on the average) 100 English ones. The trained tree was used to predict the accented variant from the German canonical one for each entry in the baseline dictionary. To be able to use the enhanced dictionary in our standard German recognition system, a mapping of those English phonemes that do not exist in German was necessary. This mapping can be found in Appendix C. Adding the non-native variants to the dictionary performed better than replacing the canonical ones. However, adding the variants means doubling the number of entries in the dictionary, which might increase the confusability. Furthermore, mapping those English phonemes that do not exist to the closest German phonemes performed better than using a merged German/English HMM model set.

8.4.1 Classification Trees

By using an English phoneme recogniser for decoding German speech, we obtained English-accented pronunciations for the German words. Since we are faced with high phoneme error rates (the phoneme error rate of the English phoneme recogniser on an English test set was 47.3%), it is not possible to simply add the generated variants to the dictionary as they are. The phoneme sequences that are due to recogniser errors should be removed. In general it is assumed that erroneous sequences will not appear as often as correct ones do (remember that for several repetitions of the German words, English-accented variants were generated), so only the correct parts should be retained. This can be achieved by inducing decision or classification trees [Ril96, Ril99].

In this section a rather informal description of classification trees is given. For a more detailed discussion on classification trees the interested reader is referred to [Kuh93, Bre84].

Classification trees consist of inner nodes, arcs connecting these nodes and finally leaves. The nodes are labelled with yes-no questions and depending on the answer one or the other arc is followed. Using appropriate training algorithms, such a tree can learn mappings from one data set to another (of course decision trees can be applied to numerous other machine learning tasks). In the case considered it is provided with the canonical German phoneme sequences and the English accented variants and is supposed to learn the mapping between the two. Thus the tree can be applied to the baseline dictionary after training to predict the accented variants. In the experiments the commercial tool C5.0 [C5] was used to grow the tree.

The tree was induced from a set of training examples consisting of attribute values, in this case the German phonemes in the vicinity of the phoneme under consideration, along with the class that the training data describes, in this case a single English target phoneme. A window moving from the right to the left is considered, each time predicting one of the target phonemes. Together with the phonemes in the vicinity of the source phoneme the last predicted phoneme is also used as an attribute to make a prediction for the following one. Each inner node of the tree is labelled with a question like 'Is the target phoneme a /{/?' or 'Is the phoneme two positions to the right a /n/?'. The arcs of the tree are labelled with answers to the questions of the originating nodes. Finally each leaf is labelled with a class to be predicted when answering all questions accordingly from the root of the tree down to that leaf, following the arcs. If the training data are non-deterministic, i.e., several classes are contained for a specific attribute constellation needed to reach a leaf, it can hold a distribution of classes seen. In this case the class that has most often been seen is predicted. After the tree is built from the training data it can be used to predict new cases by following the questions from the root node to the leaf nodes. In addition to the phoneme under consideration itself, the last predicted phoneme and a phoneme context of three to the left and three to the right was used, making up a total number of eight attributes. A problem that arose was that of different lengths of the phoneme strings. Before training the decision tree they needed to be aligned to have the same length. An iterative procedure was used that starts with those entries in the training data that have the same number of phonemes. Co-occurrences are calculated on these portions and used to continuously calculate the alignment for all other entries, by inserting so-called 'null-phonemes' at the most probable places (see [Rap98]) until the whole training set is aligned.

One of the first questions that can be found in the generated tree is the question whether the considered phoneme is a glottal stop /?/. If so it is removed in the English target output. This is very reasonable since in English the glottal stop does not exist. In Figure 8.10 another example of a set of questions is shown as a very small subtree that was generated for the German phoneme /y:/ that occurs e.g. in the word 'Büro' /by:ro:/, (Engl. 'office'). In the case where the last predicted English phoneme was a /b/ and the last two phonemes were /-/ (which is the sign representing the null-phonemes, here very probably representing a word boundary), the predicted English phoneme will be a /uː/. This makes sense, since English speakers are usually not able to pronounce the German /yː/ and it is in practice very often substituted by the phoneme /uː/. In the figure only those arcs of the subtree that were labelled with 'yes' are shown in detail. The whole tree consisted of 16,992 nodes and 18,363 leafs.

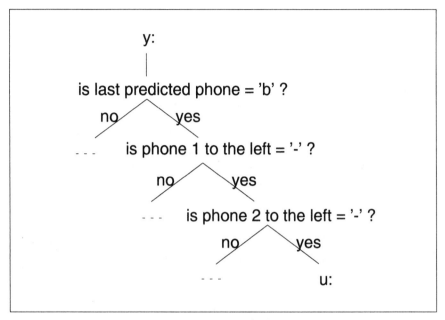

Fig. 8.10. Subtree for the German phoneme /yː/

8.4.2 Experiments and Results

A total number of 190 speakers from the German command set was used for the phoneme recognition experiments that generated the English-accented transcriptions. Eight native English speakers speaking German were excluded to build the test set for the final word recognition experiments. A set of experiments using a subset of the command corpus was conducted to find the optimal settings w.r.t. the LM weight and word insertion penalties of the phoneme recogniser. This was to avoid too many phoneme insertions, which would occur frequently if no penalty was applied. From the speech files the silence in the beginning and in the end was cut off to prevent the recogniser from hypothesising fricatives or the like, where actually silence occurred. We tried to balance insertions and deletions, however always keeping more insertions than deletions, since we expect to find some typical insertions for that source language. A decision tree was trained with the phoneme strings generated this way and applied to the German dictionary to generate English-accented German variants.

The goal was to improve the recognition results for English-accented German speech using our standard German recognition system. Thus apart from the dictionary, all settings remained unchanged. First the baseline WER results for eight native speakers of English (three American and five British English speakers) were measured and compared to the slightly changed test set of 15 native German speakers that was used in Chapter 6. The overall result on the command corpus was 11.5% WER compared to 18% WER for the non-native speakers, which corresponds to an increase in WER of 56.5% relative.

The non-native speakers were then tested using the dictionary enhanced with the variants. Figure 8.11 shows the results for the eight speakers using the baseline dictionary ('BaseDict'), the baseline dictionary combined with our weighted MLLR adaptation ('Base+MLLR'), the dictionary extended with the generated English-accented variants ('extdDict') and finally the extended dictionary combined with MLLR adaptation ('extdMLLR'). The American speakers are ID059, ID060 and ID075.

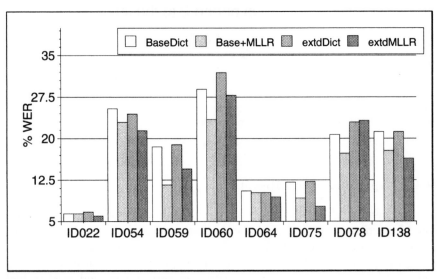

Fig. 8.11. WERs using the extended dictionary

Weighted MLLR adaptation can improve the baseline results for most of the speakers. Relative improvements of up to 37% can be achieved. The extended dictionary can only improve the results for two speakers, when no additional speaker adaptation is used. However, when combining the extended dictionary with MLLR the results can be improved for most of the speakers compared to the baseline dictionary. Further, for five out of the eight speakers the extended dictionary with MLLR is better than the baseline dictionary with MLLR and further improvements of up to 16% can be achieved. When testing the enhanced dictionary on the native reference speakers only a slight increase in WER from 11.5% to 11.8% can be observed, which might be due to the increased confusability.

More results are shown in Figure 8.12. The experiments using the extended dictionary and weighted MLLR were repeated using the semi-supervised approach (i.e. weighted MLLR combined with the CM). It can be seen that especially for those speakers with high initial WERs, the use of the semi-supervised approach is superior to the unsupervised one, because it can be avoided to use misrecognised utterances for adaptation. Improvements of up to 22% can be achieved compared to unsupervised adaptation.

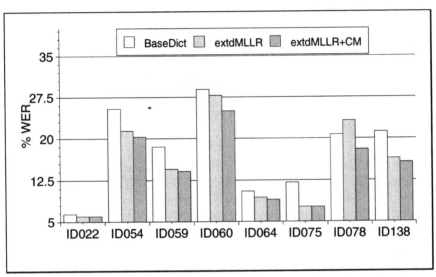

Fig. 8.12. WERs for unsupervised and semi-supervised adaptation, using the extended dictionary

The big differences in the baseline recognition rate indicate that the strength of accent varies dramatically in the test set. An auditive test revealed that while two speakers had only a very slight accent (ID022 and ID064), one (ID078) had a very strong accent. This is reflected in the baseline WERs, where ID022 and ID064 achieve the lowest WERs and ID078 is among those who achieve high WERs. Furthermore mixing native British and American speakers is also not optimal, but due to the lack of more British English speakers the American ones were used to have a more representative test set. Figure 8.13 shows again the overall results, averaged over all speakers. It can be seen that the extended dictionary combined with MLLR outperforms the baseline, however, not MLLR. But when the CM is additionally used, the weighted MLLR can be outperformed.

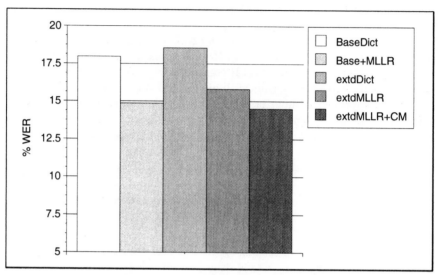

Fig. 8.13. Overall results

The way the accented pronunciations are generated does not at all consider possible differences in accents. As we have learned from the experiments on the ISLE database, a speaker-wise selection of relevant pronunciations is beneficial and sometimes even necessary to obtain improvements. Comparing these results to the ISLE experiments, simply generating one non-native variant for each entry in the dictionary that is the same for all speakers, corresponds to e.g. using all German rules to construct one dictionary for all German speakers. The results showed that this could only improve the performance for some of the speakers and was clearly outperformed by the speaker-specific dictionary that used only selected rules for each speaker.

As a consequence instead of using the decision tree directly to generate the variants as was done in the current experiments, rules should be derived from the tree, such that a speaker-wise selection becomes possible. Then improvements for more speakers are to be expected, also for the extended dictionary alone. Unfortunately at the time of the writing of this book it was not possible to do so.

The results however indicate that the improvements that can be achieved by acoustic adaptation using the weighted MLLR approach and by the proposed pronunciation adaptation method are at least partly additive.

8.5 Summary

In this section the problem of recognising non-native speech was examined and a new method for generating non-native pronunciations without using non-native data was proposed. It was shown that performance decreases if a recogniser that is optimised for native speech is exposed to non-native speech. Investigations on the ISLE database showed that including specialised non-native pronunciation variants in the dictionary can greatly improve the results. A set of rules was manually derived and the optimal set of rules was determined for each speaker in the test set. The application of these specialised rules was clearly superior to using all rules or rules that are typical for the native language of a speaker group. However, since the manual derivation is not feasible if the combination of several source and target languages is to be covered, a new method was introduced that solely relies on native speech to derive non-native variants automatically. The big advantage of the proposed method is that it requires native data only, and can thus be repeated for all the languages for which HMM models trained on native speech are available.

English native speakers who spoke German were investigated in more detail. To derive the accented pronunciation variants, German speech was decoded with an English phoneme recogniser. A decision tree was trained to map the German canonical pronunciation to the English-accented variants obtained. The trained tree was then used to generate an extended dictionary, containing the native and the non-native variants. When combining the enhanced dictionary with MLLR speaker adaptation, the results can be improved for all speakers compared to the baseline dictionary and for the majority of the speakers, compared to the baseline dictionary with MLLR adaptation. While weighted MLLR alone can achieve up to 37% reduction in WER, the additional usage of the enhanced dictionary achieves further improvements of up to 16%. The use of the semi-supervised adaptation is superior to the unsupervised approach, especially for speakers with high initial WERs. Using the enhanced dictionary with the native speakers, the performance remained almost stable.

However, it should be possible to further improve the results, if rules are generated instead of using the tree directly, so that a speaker-wise selection of the variants becomes possible. The experiments showed that the improvements that can be achieved by acoustic adaptation (using the weighted MLLR approach) and the proposed pronunciation adaptation method are at least partly additive and should therefore be combined.

The approach is very promising because of its flexibility w.r.t. the considered languages. Only native data for the source and target language under consideration is required to derive accented pronunciations of the target language spoken with the accent of the source language. Even though in the experiments only the language pair German (target language) and English (source language) was considered, the proposed method is valid for arbitrary language pairs and is able to overcome the problem of insufficient accented speech data.

When pronunciation rules are to be selected specifically for each speaker, this has so far been done by testing each rule separately. However in real applications this is not feasible and the selection of rules will preferably be conducted dynamically. For this purpose all previously generated rules could be represented in a multi-dimensional feature space, that represents different 'ways of pronunciation'. The appropriate rules for the current speaker would then be determined and chosen for dictionary adaptation. A method to conduct such a dynamic selection is currently under investigation and is briefly outlined in Chapter 9.

9 Future Work

Although it has been demonstrated that the combination of the new, dynamic, weighted MLLR approach with CMs and pronunciation adaptation yields a robust adaptation to non-native speakers, there are some topics that should be investigated further in the future.

First, it would be desirable to have larger amounts of native as well as of non-native speech to train SD systems and compare their performance with the performance achieved by the unsupervised and semi-supervised weighted MLLR approaches using the same data. It was not possible to test this, since no appropriate database was available.

Second, for pronunciation adaptation, the trained decision tree was used directly to derive English-accented variants from the German canonical ones. As we have learned from the results achieved on the ISLE data base, a speaker-wise selection of pronunciation rules is clearly preferable to simply generating variants for all entries in the dictionary. Therefore, pronunciation rules should be generated for each speaker instead of using the tree directly to enable such a speaker-wise selection of relevant rules. Once more speaker-specific pronunciations are available it might be worthwhile to consider the combination of weighted MLLR and MAP for acoustic adaptation (if enough adaptation data is available), since MAP adapts specific phonemes, which might be helpful in the case of non-native speech. In the current case many phonemes in the dictionary may be wrong, because of the rather unspecific (as far as single speakers are concerned) pronunciation modelling. Therefore applying MAP might be dangerous. Once a speaker specific pronunciation modelling is available, that allows one to be pretty sure about having correct pronunciations for each speaker, then MAP can be applied without much risk.

In order to be able to dynamically choose the relevant rules, according to the type and strength of accent of the current speaker in an online way, a new method is under investigation. This method is very briefly reviewed in the following.

9.1 Dynamic Selection of Pronunciation Rules

The description of the basic idea is rather informal and not intended to specify a detailed algorithm for solving the problem.

This new approach for the dynamic selection of pronunciation rules was inspired by the Eigenvoices approach, which is a technique for (acoustic) speaker adaptation and was briefly introduced in Chapter 6, see [Kuh99, Kuh98b, Kuh98a, Ngu99b]. While the Eigenvoice approach assumes that each speaker can be characterised by his voice, here it is assumed that each speaker can be characterised by his pronunciations in addition to his voice, speaking rate etc. Or to be more general, each accent (assuming the target language is fixed) is characterised by its mother tongue. In this approach each speaker or each accent is described or characterised by a set of pronunciation rules that transforms the native canonical pronunciation of the target language into its accented variant.

For this purpose the rules for all available source languages are represented in a n-dimensional pronunciation space. Whenever a speaker starts using the system, he is located in this space. One way to perform this localisation could be to run a phoneme recogniser in parallel to the recogniser and use the result to determine distances to the rule sets. Depending on his location in the space and based on distance measures, the closest (set of) rules can be determined. These rules can then be applied to the basic dictionary that contains the canonical pronunciation only. After the application of the rules, the dictionary will be adapted to the speaker's way of pronouncing the words. All we need to do now is to find a good representation of the rules in the pronunciation space. If we then used principal component analysis (PCA) for dimensionality reduction, this would yield, comparable to the Eigenvoices, Eigenpronunciations. However, the approach is certainly not limited to that; we could also think of employing NNs, decision trees etc. for the final selection of the rules.

A schematic view of a two-dimensional pronunciation space is depicted in Figure 9.1. It contains rule sets of the source languages English, Japanese, French and Polish. The target language is assumed to be fixed.

For the new speaker who was located in this space, the closest rules are determined and thus a speaker-specific rule set is found. By selecting separate rules instead of complete source language-specific rule sets, we can account for different strengths in accent.

In case the speaker's mother tongue is not represented by a rule set in the pronunciation space, because such a rule set is simply not available, nevertheless the closest rules can be selected. For example, the mother tongue of a speaker is Russian. Russian is not represented in our pronunciation space, Polish is. Certainly a Polish accent is somewhat different from a Russian one, but there are correspondences. And the Polish accent is certainly closer to Russian than e.g. French is. So by choosing and applying the Polish rule set we can adapt the dictionary to the Russian speaker, which should then con-

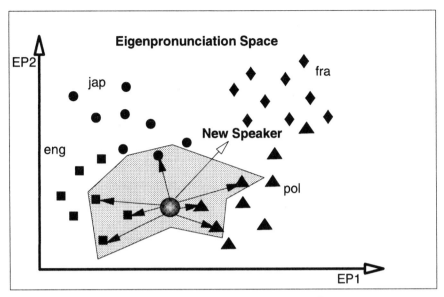

Fig. 9.1. Selecting speaker-specific pronunciation rules in the pronunciation space

tain pronunciations that are closer to the Russian speaker than the baseline dictionary was. The ISLE experiments have shown that rules from another source language (other than the mother tongue of a speaker) can also improve the results.

This new approach for dynamically selecting pronunciation rules thus has the potential to generate speaker adapted dictionaries even for unknown accents and different strengths of accents that could then be combined with semi-supervised speaker adaptation. However, a lot of research work is still necessary in this area.

10 Summary

Automatic speech recognition (ASR) systems have reached a technological level that allows their usage in real applications. Leaning towards the human processing of speech various models need to be built to reflect these procedures. Since these models perform badly when there is a mismatch between the situation the system was faced with during the training stage of the models and the situation in which it is actually used or tested, adaptation methods are widely used to adapt the models to the current situation.

The research presented in this book concentrated on the adaptation of the acoustic models to new speakers and the adaptation of the pronunciation dictionary to non-native accents. Non-native speakers pose a big problem to state-of-the-art ASR systems, since all models of these systems are usually trained exclusively on native data. However, the topic of recognising non-native speech becomes more and more important due to the growing globalisation and increasing Internet applications.

For the purpose of acoustic adaptation to new speakers the well-known method of Maximum Likelihood Linear Regression (MLLR) adaptation is extended such that it can deal with very small amounts of adaptation data. So far the standard MLLR approach could only successfully adapt the acoustic models if several sentences were available from the new speaker. For the isolated word recognition system considered here however, this was not sufficient. Using the standard MLLR approach in fact drastically increased the word error rates (WERs) for small amounts of adaptation data.

The proposed extension to this algorithm introduces two weighting schemes, yielding the 'weighted MLLR' approach. Both weighting schemes, the static and the dynamic one overcome this problem and can decrease the WERs for small amounts of adaption data. They calculate the adapted mean vector of the Gaussian densities of the HMM models as the weighted sum of the original mean as obtained from the speaker independent (SI) models (in the first adaptation step) and the 'standard MLLR-adapted' mean for all following steps, respectively, and the updated mean calculated from the adaptation data. By using this simple way of incorporating prior information, this approach can overcome the problem of over-training on sparse adaptation data and decreases the WERs by up to 24% using as few as ten seconds of adaptation data in one adaptation step. Other approaches to this problem depend on the estimation of more complex prior densities.

The method was first tested on native speakers to prove its general validity and it was shown that both weighting schemes achieve comparable improvements. However the dynamic one has proven to be much more stable than the static one. The optimal static weighting factor turned out to depend heavily on various factors, such as the number of regression classes and number of speech frames used for adaptation. Finding the optimal value needed to be repeated whenever the recogniser settings changed, thus involving a big empirical effort. Furthermore the choice of a sub-optimal static weight sometimes led to a drastic increase in WER.

In contrast the dynamic weighting factor turned out to be much more robust w.r.t. the mentioned factors. Since it takes into account the number of speech frames that have been observed so far from the current speaker, only its initial value needs to be adjusted manually. Varying this initial weight over a very wide range did not have much influence on the overall performance. Since the application of the dynamic weight is much less critical than the application of the static weight, the dynamic algorithm is clearly preferable for any kind of application. A more detailed examination of the changed model parameters furthermore revealed that the dynamic weighting makes major changes to the models at the beginning; subsequently, if more data is used, a fine tuning is done. This seems more reasonable than the static algorithm, in which each utterance has the same influence on the models, no matter how long the speaker has been using the system.

One known problem in unsupervised adaptation schemes is that of using misrecognised utterances for adaptation. This might result in an erroneous estimation of the adaptation parameters and thus in an increased WER. To overcome this problem, confidence measures (CMs) are used to judge how reliably an utterance was recognised. Only utterances that were recognised with high confidence are used for adaptation. In this research work a neural network (NN) classifier was employed that used a set of features that were extracted during recognition and from the recogniser output. In addition to well-known features that are mostly related to the acoustic score, some new ones have been introduced. These are based on phoneme durations and were motivated by the fact that often phoneme durations that are determined by the segmentation of the recogniser deviate severely from the corresponding duration found in the training database in case of a misrecognition. By combining the traditional features with the newly derived ones the classification rate of the NN could be improved from 91% to 94%, while simultaneously identifying 95% of out of vocabulary (OOV) words. Also separately testing the new features delivered satisfying results, though these results were slightly worse than those obtained using the traditional features if tested separately. But whereas the duration-based features need only the first-best hypothesis to be computed, the traditional features require the computation of an N-best list, which might be problematic for some applications.

When combined with weighted MLLR, this semi-supervised approach outperformed the unsupervised one, especially if a large number of OOV words were included in the test set.

Especially for the non-native speakers that were then considered in more detail, such a semi-supervised approach is suitable since usually the initial WERs are much higher compared to native speakers. Indeed an increase in WER could be observed on the non-native test set compared to the native test set. The WERs could be decreased by up to 37% using weighted MLLR and further improvements were possible with the semi-supervised approach.

In this special case of non-native speakers the acoustic adaptation alone is not considered sufficient, since apart from the difference in the acoustic realisation of the sounds, deviations in the way the words are pronounced will also be present. As a consequence the adaptation of the pronunciation dictionary is considered to be important. The necessity of explicitly modelling non-native pronunciation was shown in experiments conducted on the ISLE database. Non-native pronunciation rules were obtained manually from the available transcriptions and their influence when applied to the baseline dictionary was tested. The experiments showed that a speaker-specific selection of rules and thus the construction of speaker specific dictionaries could improve performance for almost all speakers, while the application of a whole set of rules that is considered typical for speakers of a certain mother tongue failed for many speakers. Combination with speaker adaptation showed the improvements of both methods to be additive.

Since a flexible approach that is applicable to any language spoken with any accent is desired, the manual derivation of pronunciation rules is not feasible. As a consequence a new method to derive non-native pronunciation variants was proposed that relies on native data bases of the considered languages only. This is in contrast to known approaches, that either require linguistic knowledge or accented databases, both of which are costly to obtain.

The case considered was that of English native speakers speaking German. An English phoneme recogniser was used to recognise German speech. The resulting English phoneme transcriptions for the German words were used to induce a decision tree, that learned the mapping from the German canonical pronunciation to the corresponding English-accented variant. Using the dictionary enhanced with these accented variants and combining this with the unsupervised weighted MLLR approach, WER reductions of up to 16% could be achieved for some speakers compared to using the unsupervised weighted MLLR alone. Using the semi-supervised approach further improvements of up to 22% were possible.

The proposed pronunciation adaptation method was tested on one language pair. However, it can be used for any other combination of languages for which native speech data are available. Since large amounts of native speech data are much easier to obtain than the corresponding amounts of non-native data, this approach is much more flexible than traditional ones. Since there is some potential to further improve the proposed method, even better results can be expected for the future.

Finally, a method for a possible dynamic, online and speaker-specific selection of pronunciation rules was informally described.

Bibliography

[AD99] M. Adda-Decker and L. Lamel. Pronunciation Variation across System Configuration, Language and Speaking Style. *Speech Communication*, 29:83–98, 1999.

[Ama00] I. Amadal, F. Korkmazskiy, and A. C. Suredan. Data-driven Pronunciation Modelling for Non-Native Speakers using Association Strength between Phones. In ASR2000, Paris, France [ASR00], pages 85–90.

[Ana98] T. Anastsakos and S. V. Balakrishnan. The use of CM in unsupervised adaptation of SRs. In ICSLP98, Sydney, Australia [ICS98], pages 2303–2306.

[ASR97] IEEE. *Automatic Speech Recognition and Understanding*, 1997.

[ASR99] IEEE. *Automatic Speech Recognition and Understanding*, 1999.

[ASR00] ISCA. *Automatic Speech Recognition*, volume 1, 2000.

[Bar98] J. Barker, G. Williams, and S. Renals. Acoustic Confidence Measures for Segmenting Broadcast News. In ICSLP98, Sydney, Australia [ICS98], pages 2719–2722.

[Bau72] L. E. Baum. An Inequality and Associated Maximization Technique in Statistical Estimation for Probabilistic Functions of Markov Processes. *Inequalities*, 3:1–8, 1972.

[Ben98] M. C. Benitez, A. Rubio, P. Garcia, and J. D. Verdejo. Word Verification using Confidence Measures in Speech Recognition. In ICSLP98, Sydney, Australia [ICS98], pages 771–774.

[Ber97] E. Bernstein and W. R. Evans. OOV Utterance Detection based on the recognizer response function. In Eurospeech97, Rhodes, Greece [Eur97].

[Ber98] G. Bernadis and H. Bourlard. Improving posterior based CM in Hybrid HMM/ANN SR Systems. In ICSLP98, Sydney, Australia [ICS98], pages 775–778.

[Bon98] P. Bonaventura, F. Gallocchio, J. Mari, and G. Micca. Speech Recognition Methods for Non-Native Pronunciation Variations. In ROLDUC98 [ROL98], pages 7–22.

[Bot00] H. Botterweck. Very fast adaptation for Large Vocabulary Continuous Speech Recognition Using Eigenvoices. In ICSLP2000, Beijing, China [ICS00], pages 334–337.

[Bou99] G. Bouwman, J. Sturm, and L. Boves. Incorporating Confidence Measures in the Dutch Train Time Table Information System Developed in the Arise Project. In ICASSP99, Phoenix, USA [ICA99], pages 493–496.

[Bre84] L. Breiman, J.H. Friedman, R. A. Olsen, and C. J. Stone. *Classification and Regression Trees*. Wadsworth & Brooks, 1984.

[Byr97] B. Byrne, M. Finke, S. Khudanpur, J. McDonough, H. Nock, M. Riley, M. Saraclar, C. Wooters, and G. Zavaliagkos. Pronunciation Modelling for Conversational Speech Recognition: A Status Report from WS97. In ASRU97, Santa Barbara, USA [ASR97], pages 26–33.

[Byr99] W. Byrne and A. Gunawardana. Discounted Likelihood Linear Regression for Rapid Adaptation. In Eurospeech99, Budapest, Hungary [Eur99], pages 203–206.

[C5] http://www.rulequest.com.

[Cam97] J. Caminero, L. Hernandez, C. de la Torre, and C. Martin. Improving Utterance Verification Using Hierarchical Confidence Measures in Continuous Natural Numbers Recognition. In ICASSP97, Munich, Germany [ICA97], pages 891–894.

[Car86] D. W. Carroll. *Psychology of Language*. Brooks/Cole, 1986.

[CEL] http://www.kun.nl/celex/.

[CG95] F. J. Caminero-Gil, C. de la Torre, L. A. Hernandez-Gomez, and C. Martin del Alamo. New n-best Based Rejection Techniques for Improving a Real-Time Telephonic Connected Word Recognition System. In Eurospeech95, Madrid, Spain [Eur95], pages 2099–2102.

[Cha96] L. Chase. Word and Acoustic Confidence Annotation for LVCSR. In ICSLP96, Philadelphia, USA [ICS96].

[Cha97] L. Chase. *Error-Responsive Feedback Mechanisms for Speech Recognisers*. PhD thesis, Carnegie Mellon University, Pittsburgh, USA, 1997.

[Cha01] D. Charlet. Confidence-Measure-Driven Incremental Adaptation for HMM-Based Speech Recognition. In ICASSP2001, Salt Lake City, USA [ICA01], pages 357–360.

[Che98] B. Chen, H.-M. Wang, L.-F. Chien, and L.-S. Lee. A*-admissible Key-Phrase Spotting with Sub-Syllable Level Utterance Verification. In ICSLP98, Sydney, Australia [ICS98], pages 783–786.

[Che99] C. Chesta, O. Siohan, and C.-H. Lee. Maximum A Posteriori Linear Regression for HMM Adaptation. In Eurospeech99, Budapest, Hungary [Eur99], pages 211–214.

[Chi97] J.-T. Chien, C.-H. Lee, and H.-C. Wang. Improved Bayesian Learning of Hidden Markov Models for Speaker Adaptation. In ICASSP97, Munich, Germany [ICA97], pages 1027–1030.

[Cho99] W. Chou. Maximum A Posteriori Linear Regression With Elliptically Symmetric Matrix Variate Priors. In Eurospeech99, Budapest, Hungary [Eur99], pages 1–4.

[Cla98] T. Claes, I. Dologlou, L. ten Bosch, and van D. Compernolle. A Novel Feature Transformation Vocal Tract Length Normalization in Automatic Speech Recognition. *Transactions on Speech, Audio Processing*, 6:549–557, November 1998.

[Col95] D. Colton, M. Fanty, and R. Cole. Utterance Verification Improves Closed-Set Recognition and Out-Of-Vocabulary Rejection. In Eurospeech95, Madrid, Spain [Eur95], pages 1067–1070.

[Cox96] S. Cox and R. Rose. Confidence Measures for the SWITCHBOARD database. In ICASSP96, Atlanta, USA [ICA96], pages 511–514.

[Cre97] N. Cremelie and J.-P. Martens. Automatic Rule-Based Generation Of Word Pronunciation Networks. In Eurospeech97, Rhodes, Greece [Eur97], pages 2459–2462.

[Cre99] N. Cremelie and J.-P. Martens. In Search of Better Pronunciation Models for Speech Recognition. *Speech Communication*, 29:115–136, 1999.

[Cry97] D. Crystal. *A Dictionary of Linguistics and Phonetics*. Blackwell, fourth edition, 1997.

[Des96] N. Deshmukh, M. Weber, and J. Picone. Automated Generation Of N-Best Pronunciations Of Proper Nouns. In ICASSP96, Atlanta, USA [ICA96], pages 283–286.

[Dig95a] V. Digalakis and L. Neumeyer. Speaker Adaptation Using Combined Transformation and Bayesian Methods. In ICASSP95, Detroit, USA [ICA95], pages 680–683.

[Dig95b] V. Digalakis, D. Rtischev, and L. Neumeyer. Speaker Adaptation using constrained estimation of Gaussian mixtures. *Transactions on Speech, Audio Processing*, 3:357–365, 1995.

[dM98] Renato di Mori, editor. *Spoken Dialogues with Computers*. Academic Press, 1998.

[Dol98] J. G. A. Dolfing and A. Wendemuth. Combination of CMs in Isolated Word Recognition. In ICSLP98, Sydney, Australia [ICS98], pages 3237–3240.

[Dow98] S. Downey and R. Wiseman. Dynamic and Static Improvements to Lexical Baseforms. In ROLDUC98 [ROL98], pages 157–162.

[Dud] http://www.duden.de.

[Eid95] E. Eide, H. Gish, P. Jeanrenaud, and A. Mielke. Understanding and Improving Speech Recognition Performance through the Use of Diagnostic Tools. In ICASSP95, Detroit, USA [ICA95], pages 221–224.

[Eid99] E. Eide. Automatic Modeling of Pronunciation Variations. In Eurospeech99, Budapest, Hungary [Eur99], pages 451–454.

[Eur95] European Speech Communication Association (ESCA). *4th European Conference on Speech Communication and Technology*, 1995.

[Eur97] European Speech Communication Association (ESCA). *5th European Conference on Speech Communication and Technology*, 1997.

[Eur99] European Speech Communication Association (ESCA). *6th European Conference on Speech Communication and Technology*, 1999.

[Eur01] International Speech Communication Association (ISCA). *7th European Conference on Speech Communication and Technology*, 2001.

[Eve01] G. Evermann and P. C. Woodland. Large Vocabulary Decoding and Confidence Estimation Using Word Posterior Probabilities. In ICASSP2001, Salt Lake City, USA [ICA01].

[Fal99] R. Falthauser, T. Pfau, and G. Ruske. Creating Hidden Markov Models for Fast Speech by Optimized Clustering. In Eurospeech99, Budapest, Hungary [Eur99], pages 407–410.

[Fen95] M.-W. Feng. Speaker Adaptation Based on Spectral Normalization and Dynamic HMM Parameter Adaptation. In ICASSP95, Detroit, USA [ICA95], pages 704–707.

[Fet96] P. Fetter, F. Dandurand, and P. Regel-Brietzmann. Word graph rescoring using Confidence Measures. In ICSLP96, Philadelphia, USA [ICS96], pages 10–13.

[Fin97] M. Finke and A. Waibel. Speaking Mode Dependent Pronunciation Modelling in Large Vocabulary Conversational Speech Recognition. In Eurospeech97, Rhodes, Greece [Eur97], pages 2379–2382.

[Fis98] A. Fischer, Y. Gao, and E. Janke. Speaker Independent Upfront Dialect Adaptation in A Large Vocabulary Continuous Speech Recognizer. In ICSLP98, Sydney, Australia [ICS98], pages 787–790.

[Fis99] A. Fischer and V. Stahl. Database And Online Adaptation For Improved Speech Recognition In Car Environments. In ICASSP99, Phoenix, USA [ICA99], pages 445–448.

[FL99] E. Fosler-Lussier and N. Morgan. Effects of Speaking Rate and Word Frequency on Pronunciations in Conversational Speech. *Speech Communication*, 29:137–158, 1999.

[Fra99] H. Franco, L. Neumeyer, M. Ramos, and H. Bratt. Automatic Detection of Phone-Level Mispronunciation for Language Learning. In Eurospeech99, Budapest, Hungary [Eur99], pages 851–854.

[Fuk97] T. Fukada and J. Sagisaka. Automatic Generation of a Pronunciation Dictionary Based on a Pronunciation Network. In Eurospeech97, Rhodes, Greece [Eur97], pages 2471–2474.

[Fuk98] T. Fukada, T. Yoshimura, and J. Sagisaka. Automatic Generation of Multiple Pronunciations Based on Neural Networks and Language Statistics. In ROLDUC98 [ROL98], pages 41–46.

[Fur01] Sadaoki Furui. *Digital Speech Processing, Synthesis and Recognition*. Marcel Dekker, Inc., second edition, 2001.

[Gal96a] M.F.J. Gales. The generation and use of regression trees for MLLR adaptation. Technical report, University of Cambridge, 1996.

[Gal96b] M.F.J. Gales and P.C. Woodland. Mean and Variance Adaptation within the MLLR Framework. *Computer Speech and Language*, 10:249–264, 1996.

[Gao00] Y. Gao, B.Ramabhadran, and M. Picheny. New Adaptation Techniques for Large Vocabulary Continuous Speech Recognition. In ASR2000, Paris, France [ASR00], pages 107–111.

[GM99] C. Garcia-Mateo, W. Reichl, and S. Orthmanns. On Combining Confidence Measures in HMM-Based Speech Recognizers. In ASRU99, Colorado, USA [ASR99], pages 201–104.

[Gor99] S. Goronzy and R. Kompe. A MAP-like weighting scheme for MLLR speaker adaptation. In Eurospeech99, Budapest, Hungary [Eur99], pages 5–8.

[Gor00a] S. Goronzy, K. Marasek, A. Haag, and R. Kompe. Phone-Duration-Based Confidence Measures for Embedded Applications. In ICSLP2000, Beijing, China [ICS00], pages 500–503.

[Gor00b] S. Goronzy, K. Marasek, R. Kompe, and A. Haag. Prosodically Motivated Features for Confidence Measures. In ASR2000, Paris, France [ASR00], pages 207–212.

[Gor01a] S. Goronzy, R. Kompe, and S. Rapp. Generating Non-Native Pronunciation Variants for Lexicon Adaptation. In *Adaptation Methods for Speech Recognition*, volume 1, pages 143–146. ISCA, 2001.

[Gor01b] S. Goronzy, M. Sahakyan, and W. Wokurek. Is Non-Native Pronunciation Modelling Necessary? In Eurospeech2001, Aalborg, Denmark [Eur01], pages 309–312.

[Gre99] S. Greenberg. Speaking in shorthand - A syllable-centric perspective for understanding Pronunciation Variation. *Speech Communication*, 29:159–176, 1999.

[Gun98] A. Gunawardana, H.-W. Hon, and L. Jiang. Word-based Acoustic CMs for LVSR. In ICSLP98, Sydney, Australia [ICS98], pages 791–794.

[Gup98] S. K. Gupta and K. Soong. Improved Utterance Rejection using Length Dependent Thresholds. In ICSLP98, Sydney, Australia [ICS98], pages 795–798.

[Han99a] P. Hanna, D. Stewart, and J. Ming. The Application of an Improved DP Match for Automatic Lexicon Generation. In Eurospeech99, Budapest, Hungary [Eur99], pages 475–478.

[Han99b] P. Hanna, D. Stewart, J. Ming, and F. J. Smith. An Improved Method of Generating Alternative Lexicon Pronunciations. In *1999 International Congress on Phonetic Science*, volume 3, pages 1713–1716. IEEE, 1999.

[He01] X. He and Y. Zhao. Model Complexity Optimization for Nonnative English Speakers. In Eurospeech2001, Aalborg, Denmark [Eur01], pages 1461–1463.

[Hei98] H. Heine, G. Evermann, and U. Jost. An HMM-Based Probabilistic Lexicon. In ROLDUC98 [ROL98], pages 57–62.

[Her99] D. Herron, W. Menzel, E. Atwell, R. Bisiani, F. Daneluzzi, R. Morton, and J. A. Schmidt. Automatic Localization and Diagnosis of Pronunciation Errors for Second-Language Learners of English. In Eurospeech99, Budapest, Hungary [Eur99], pages 855–858.

[Hol99] T. Holter and T. Svendsen. Maximum Likelihood Modelling of Pronunciation Variation. *Speech Communication*, 29:177–191, 1999.

[Hom97] S. Homma, K. Aikawa, and S. Sagayama. Improved Estimation of Supervision in Unsupervised Speaker Adaptation. In ICASSP97, Munich, Germany [ICA97], pages 1023–1026.

[Hop85] Hopfield and Tank. *"Neural" Computation of Decisions in Optimization Problems*. Springer Verlag, 1985.

[HTK] http://htk.eng.cam.ac.uk.

[Hua00] C. Huang, E. Chang, J. Zhou, and K.-F. Lee. Accent Modeling Based on Pronunciation Dictionary Adaptation for Large Vocabulary Mandarin Speech Recognition. In ICSLP2000, Beijing, China [ICS00], pages 818–821.

[Hum96] J. J. Humphries, P. C. Woodland, and D. Pearce. Using Accent-Specific Pronunciation Modelling for Robust Speech Recognition. In ICSLP96, Philadelphia, USA [ICS96], pages 2324–2327.

[Huo95] Q. Huo and C. Chan. On-line Bayes Adaptation of SCHMM Parameters for Speech Recognition. In ICASSP95, Detroit, USA [ICA95], pages 708–711.

[Huo96] Q. Huo and C.-H. Lee. A Study of On-line Quasi-Bayes Adaptation For CDHMM-Based Speech Recognition. In ICASSP96, Atlanta, USA [ICA96], pages 705–709.

[Hwa97] T.-H. Hwang, L.-M. Lee, and H.-C. Wang. Feature Adaptation Using Deviation Vector For Robust Speech Recognition in Noisy Environment. In ICASSP97, Munich, Germany [ICA97], pages 1227–1230.

[ICA95] IEEE. *1995 International Conference on Acoustics, Speech and Signal Processing*, 1995.

[ICA96] IEEE. *1996 International Conference on Acoustics, Speech and Signal Processing*, 1996.

[ICA97] IEEE. *1997 International Conference on Acoustics, Speech and Signal Processing*, 1997.

[ICA98] IEEE. *1998 International Conference on Acoustics, Speech and Signal Processing*, 1998.

[ICA99] IEEE. *1999 International Conference on Acoustics, Speech and Signal Processing*, 1999.

[ICA01] IEEE. *2001 International Conference on Acoustics, Speech and Signal Processing*, 2001.

[ICS96] *International Conference on Spoken Language Processing*, 1996.

[ICS98] *International Conference on Spoken, Language Processing*, 1998.

[ICS00] *International Conference on Spoken Language Processing*, 2000.

[Ida98] M. Ida and R. Yamasaki. An Evaluation of KWS Performance Utilizing False Alarm Rejection Based on Prosodic Information. In ICSLP98, Sydney, Australia [ICS98], pages 803–806.

[Ins98] Institute of Parallel and Distributed High Performance Systems, University of Tübingen, Department of Computer Architecture. *SNNS, Stuttgart Neural Network Simulator, User Manual v4.2*, 1998. http://www-ra.informatik.uni-tuebingen.de/SNNS.

[IPA] http://www2.arts.gla.ac.uk/ipa/ipa.html.

[ISL] http://nats-www.informatik.uni-hamburg.de/~isle/.

[Jia98] L. Jiang and X. Huang. Vocabulary-Independent Word CM Using Subword Features. In ICSLP98, Sydney, Australia [ICS98], pages 3245–3248.

[Jit98] T. Jitsuhiro, S. Takahashi, and K. Aikawa. Rejection of Out-of-Vocabulary Words Using Phoneme Confidence Likelihood. In ICASSP98 [ICA98], pages 217–220.

[Jos97] U. Jost, H. Heine, and G. Evermann. What Is Wrong With The Lexicon - An Attempt To Model Pronunciations Probabilistically. In Eurospeech97, Rhodes, Greece [Eur97], pages 2475–2479.

[Jun97] J. Junkawitsch, G. Ruske, and H. Höge. Efficient Methods for Detecting Keywords in Continuous Speech. In Eurospeech97, Rhodes, Greece [Eur97], pages 259–262.

[Jun98] J. Junkawitsch and H. Höge. Keyword Verification Considering the Correlation of Succeeding Feature Vectors. In ICASSP98 [ICA98], pages 221–224.

[Jur01] D. Jurafsky, W. Ward, Z. Jianping, K. Herold, Y. Xiuyang, and Z. Sen. What Kind of Pronunciation Variation is Hard for Triphones to Model? In ICASSP2001, Salt Lake City, USA [ICA01], pages 577–580.

[Kaw97] T. Kawahara, C.-H. Lee, and B.-H. Juang. Combining Key-Phrase Detection and Subword-based Verification for Flexible Speech Understanding. In ICASSP97, Munich, Germany [ICA97], pages 1159–1162.

[Kaw98a] T. Kawahara, K. Ishizuka, S. Doshita, and C.-H. Lee. Speaking-Style Dependent Lexicalized Filler Model for Key-Phrase Detection and Verification. In ICSLP98, Sydney, Australia [ICS98], pages 3253–3256.

[Kaw98b] T. Kawahara, C.-H. Lee, and B.-H. Juang. Flexible Speech Understanding based on Combined Key-Phrase Detection and Verification. *Transactions on Speech and Audio Processing*, pages 558–568, November 1998.

[Kem97] T. Kemp and T. Schaaf. Estimating Confidence Using Word Lattices. In Eurospeech97, Rhodes, Greece [Eur97], pages 827–830.

[Kem99] T. Kemp. *Ein automatisches Indexierungssystem für Fernsehnachrichtensendungen*. PhD thesis, University of Karlsruhe, December 1999. Shaker.

[Kes99] J. M. Kessens, M. Wester, and H. Strik. Improving the Performance of a Dutch CSR by Modeling Within-word and Cross-word Pronunciation Variants. *Speech Communication*, 29:193–207, 1999.

[Kie96] Kiesling. *Extraktion und Klassifikation prosodischer Merkmale in der automatischen Sprachverarbeitung*. PhD thesis, University of Erlangen, 1996. Shaker.

[Kip96] A. Kipp, M.-A. Wesenick, and F. Schiel. Automatic Detection and Segmentation of Pronunciation Variants in German Speech Corpora. In ICSLP96, Philadelphia, USA [ICS96], pages 106–109.

[Kip97] A. Kipp, M.-A. Wesenick, and F. Schiel. Pronunciation Modeling Applied To Automatic Segmentation Of Spontaneous Speech. In Eurospeech97, Rhodes, Greece [Eur97], pages 1023–1026.

[Kom97] Ralf Kompe. *Prosody in Speech Understanding Systems*. Springer Verlag, 1997.

[Koo97] M.-W. Koo, C.-H. Lee, and B.-H. Juang. A New Hybrid Decoding Algorithm for SR and Utterance Verification. In *IEEE Workshop on Speech Recognition and Understanding*. IEEE, 1997.

[Koo98] M.-W. Koo, C.-H. Lee, and B.-H. Juang. A New Decoder Based on a Generalized Confidence Score. In ICASSP98 [ICA98], pages 213–217.

[Kuh93] R. Kuhn. *Keyword Classification Trees for Speech Understanding Systems*. PhD thesis, McGill University, Montreal, October 1993.

[Kuh98a] R. Kuhn, P. Nguyen, J.-C. Junqua, and L. Goldwasser. Eigenfaces, Eigenvoices: Dimensionality reduction for specialized pattern Reduction. In *IEEE Second Workshop on Multimedia Signal Processing*, pages 71–76, 1998.

[Kuh98b] R. Kuhn, P. Ngyen, J.-C. Junqua, L. Goldwasser, N. Niedyielski, Fincke, Field, and Contolini. Eigenvoices for Speaker Adaptation. In ICSLP98, Sydney, Australia [ICS98], pages 1771–1774.

[Kuh99] R. Kuhn, P. Nguyen, J.-C. Junqua, R. Boman, N. Niedzielski, M. Fincke, K. Field, and M. Contolini. Fast Speaker Adaptation Using a priori Knowledge. In ICASSP99, Phoenix, USA [ICA99], pages 749–752.

[Lad93] P. Ladefoged. *A Course in Phonetics*. Harcourt Brace, third edition, 1993.

[Lam96] L. Lamel and G. Adda. On Designing Pronunciation Lexicons for Large Vocabulary Continuous Speech Recognition. In ICSLP96, Philadelphia, USA [ICS96], pages 6–9.

[LC99] R. Lopez-Cozar, A.-J.Rubio, P. Garcia, and J.-C. Segura. A new Word-Confidence Threshold Technique to Enhance the Performance of Spoken Dialogue Systems. In Eurospeech99, Budapest, Hungary [Eur99], pages 1395–1398.

[Lee97] C.-H. Lee. Adaptive Compensation for Robust Speech Recognition. In ASRU97, Santa Barbara, USA [ASR97], pages 357–364.

[Leg95a] C. J. Leggetter. *Improved Acoustic Modeling for HMMs using Linear Transformations*. PhD thesis, University of Cambridge, 1995.

[Leg95b] C. J. Leggetter and P. C. Woodland. MLLR for Speaker Adaptation of CDHMMs. *Computer Speech and Language*, 9:171–185, 1995.

[Leg95c] C. J. Leggetter and P. C. Woodland. Speaker Adaptation of HMMs using linear regression. Technical report, University of Cambridge, 1995.

[Let98] G. Lethinen and S. Safra. Generation and Selection of Pronunciation Variants for a Flexible Word Recognizer. In ROLDUC98 [ROL98], pages 67–70.

[Lin98] Q. L. Lin, S. Das, D. Lubensky, and M. Picheny. A New CM Based on Rank-Ordering Subphone Scores. In ICSLP98, Sydney, Australia [ICS98], pages 3249–3252.

[Lle96a] E. Lleida and R. C. Rose. Efficient Decoding and Training Procedures for Utterance Verification in Continuous Speech Recognition. In ICASSP96, Atlanta, USA [ICA96], pages 507–510.

[Lle96b] E. Lleida and R. C. Rose. Likelihood ratio decoding and confidence measures for CSR. In ICSLP96, Philadelphia, USA [ICS96], pages 478–491.

[Mar97] K. L. Markey and W. Ward. Lexical Tuning Based On Triphone Confidence Estimation. In Eurospeech97, Rhodes, Greece [Eur97], pages 2479–2482.

[Mat98] T. Matsui and S. Furui. N-best-based Unsupervised Speaker Adaptation for SR. *Computer Speech and Language*, 12:41–50, 1998.

[MB98] F. Mouria-Beji. Context and Speed Dependent Phonemic Models for CSR. In ROLDUC98 [ROL98], pages 79–84.

[McA98] D. McAllaster, L. Gillick, F. Scattone, and M. Newman. Studies with Fabricated Switchboard Data: Exploring Sources of Model-Data Mismatch. In *Proceedings of the DARPA Workshop on Conversational Speech Recognition, Hub-5*, volume 1, 1998.

[Mod97] P. Modi and M. Rahim. Discriminative Utterance Verification Using Multiple Confidence Measures. In Eurospeech97, Rhodes, Greece [Eur97], pages 103–106.

[Mok98] H. Mokbel and D. Jouvet. Derivation of the Optimal Phonetic Transcription Set for a Word form its Acoustic Realisations. In ROLDUC98 [ROL98], pages 73–78.

[Mor97] M. Morishima, T. Isobe, and J.-I. Takahashi. Phonetically Adaptive Cepstrum Mean Normalization for Acoustic Mismatch Compensation. In ASRU97, Santa Barbara, USA [ASR97], pages 436–441.

[Mor00] N. Moreau, D. Charlet, and D. Jouvet. Confidence Measure in Incremental Adaptation for the Rejection of Incorrect Data. In *2000 International Conference on Acoustics, Speech and Signal Processing*, volume 3, pages 1807–1810. IEEE, 2000.

[Net97] C. V. Neti, S. Roukos, and E. Eide. Word-based Confidence Measures as a Guide for Stack Search in Speech Recognition. In ICASSP97, Munich, Germany [ICA97], pages 883–886.

[Neu95] L. Neumeyer, A. Sankar, and V. Digalakis. A Comparative Study of Speaker Adaptation Techniques. In Eurospeech95, Madrid, Spain [Eur95], pages 1127–1130.

[Ngu99a] P. Nguyen, P. Gelin, J.-C. Junqua, and J.-T. Chien. N-Best Based Supervised and Unsupervised Adaptation for Native and Non-Native Speakers in Cars. In ICASSP99, Phoenix, USA [ICA99], pages 173–176.

[Ngu99b] P. Nguyen, C. Wellekens, and J.-C. Junqua. Maximum Likelihood Eigenspace and MLLR for speech recognition in noisy environments. In Eurospeech99, Budapest, Hungary [Eur99], pages 2519–2522.

[Noc98] H. J. Nock and S. J. Young. Detecting and Correcting Poor Pronunciations for Multiword Units. In ROLDUC98 [ROL98], pages 85–90.

[Pao98] C. Pao, P. Schmid, and J. Glass. Confidence Scoring for Speech Understanding Systems. In ICSLP98, Sydney, Australia [ICS98], pages 815–818.

[Pau98] Erwin Paulus. *Sprachsignalverarbeitung - Analyse, Synthese, Erkennung.* Spektrum Akademischer Verlag, 1998.

[Pfa97] T. Pfau, M. Beham, W. Reichl, and G. Ruske. Creating Large Subword Units for Speech Recognition. In Eurospeech97, Rhodes, Greece [Eur97], pages 1191–1194.

[Pfa98] T. Pfau and G. Ruske. Creating Hidden Markov Models for Fast Speech. In ICSLP98, Sydney, Australia [ICS98].

[Pit00] M. Pitz, F. Wessel, and H. Ney. Improved MLLR Speaker Adaptation Using Confidence Measures for Conversational Speech Recognition. In ICSLP2000, Beijing, China [ICS00], pages 548–551.

[Pol98] T. S. Polzin and A. Waibel. Pronunciation Variation in Emotional Speech. In ROLDUC98 [ROL98], pages 103–107.

[Pou97] L. Pousse and G. Perennou. Dealing With Pronunciation Variants at the Language Model Level for the Continuous Automatic Speech Recognition of French. In Eurospeech97, Rhodes, Greece [Eur97], pages 2727–2730.

[Pye97] D. Pye and P. C. Woodland. Experiments in Speaker Normalisation and Adaptation for Large Vocabulary Speech Recognition. In ICASSP97, Munich, Germany [ICA97], pages 1047–1050.

[Rab86] L.R. Rabiner and B.H. Juang. An introduction to hidden markov models. *IEEE ASSP Magazine*, pages pp. 4–16, January 1986.

[Rab89] Rabiner. A tutorial on hidden markov models and selected applications in speech recognition. *Proceedings of the IEEE*, 77 2:257–285, 1989.

[Rab93] L.R. Rabiner and B.H. Juang. *Fundamentals of Speech Recognition.* Prentice Hall, 1993.

[Rah97] M. Rahim. A Parallel Environment Model (PEM) for Speech Recognition and Adaptation. In Eurospeech97, Rhodes, Greece [Eur97], pages 1087–1090.

[Ram98] P. Ramesh, C.-H. Lee, and B.-H. Juang. Context Dependent Anti-Subword Modeling fo UV. In ICSLP98, Sydney, Australia [ICS98], pages 3233–3236.

[Rap98] S. Rapp, S. Goronzy, R. Kompe, and P. Buchner. A Speech Unit for i.LINK. In *Technical Digests of the Sony Research Forum 98*, volume 1, page 274, 1998.

[Rav97] M. Ravishankar and M. Eskenazi. Automatic Generation Of Context-Dependent Pronunciations. In Eurospeech97, Rhodes, Greece [Eur97], pages 2467–2470.

[Rey97] Reyelt. *Experimentelle Untersuchungen zur Festlegung und Konsistenz suprasegmentaler Einheiten für die automatische Spracherkennung.* PhD thesis, Technical University of Braunschweig, 1997. Shaker.

[Ril96] M. Riley and A. Ljolje. *Automatic Speech and Speaker Recognition,* volume 1 of *VLSI, Computer Architecture and Digital Signal Processing,* chapter Automatic Generation of Detailed Pronunciation Lexicons, pages 285–301. Kluwer Academic Press, 1996.

[Ril99] M. Riley, W. Byrne, M. Finke, S. Khudanpur, A. Ljolie, J. McDonough, H. Nock, M. Saraclar, C. Wooters, and G. Zavaliagkos. Stochastic Pronunciation Modelling from hand-labelled phonetic Corpora. *Speech Communication,* 29:209–224, 1999.

[Rit92] Ritter, Martinez, and Schulten. *Neuronale Netze.* Addison-Wesley, 1992.

[Riv95] Z. Rivlin. A Confidence Measure for Acoustic Likelihood Scores. In Eurospeech95, Madrid, Spain [Eur95], pages 523–526.

[Riv96] Z. Rivlin, M. Cohen, V. Abrash, and T. Chung. A Phone-Dependent Confidence Measure for Utterance Rejection. In ICASSP96, Atlanta, USA [ICA96], pages 515–519.

[ROL98] ESCA. *Workshop on Modeling Pronunciation Variation,* 1998.

[Ros98] R. C. Rose, H. Yao, G. Riccardi, and J. Wright. Integration of Utterance Verification with Statistical Language Modeling and Spoken Language Understanding. In ICASSP98 [ICA98], pages 237–240.

[Rot99] J. Rottland, D. Willett, and G. Rigoll. Speaker Adaptation Using Feature Space Transformation and HMM Adaptation for Hybrid MMI-NN/HMM Speech Recognition. In Eurospeech99, Budapest, Hungary [Eur99], pages 219–222.

[Rue97] B. Rueber. Obtaining Confidence Measures From Sentence Probabilities. In Eurospeech97, Rhodes, Greece [Eur97], pages 739–742.

[Rus92] G. Ruske, B. Plannerer, and T. Schultz. Stoachstic Modelling of Syllable-Based Units for Continuous Speech Recognition. In *International Conference on Spoken Language Processing,* volume 2, pages 1503–1506, 1992.

[Sah01] M. Sahakyan. Variantenlexikon italienischer und deutscher Lerner des Englischen für die automatische Spracherkennung. Master's thesis, IMS, University of Stuttgart, May 2001.

[SAM] http://www.phon.ucl.ac.uk/home/sampa/home.htm.

[San96] A. Sankar, L. Neumeyer, and M. Weintraub. An Experimental Study of Acoustic Adaptation Algorithms. In ICASSP96, Atlanta, USA [ICA96], pages 713–716.

[Sar99] M. Saraclar, H. Nock, and S. Khudanpur. Pronunciation Modeling by Sharing Gaussian Densities across Phonetic Models. In Eurospeech99, Budapest, Hungary [Eur99], pages 515–518.

[Sch97] T. Schaaf and T. Kemp. Confidence Measures for Spontaneous Speech. In ICASSP97, Munich, Germany [ICA97], pages 875–879.

[Set96] A. R. Setlur, R. A. Sukkar, and J. Jacob. Correcting recognition errors via discriminative utterance verification. In ICSLP96, Philadelphia, USA [ICS96], pages 602–605.

[Shi97] K. Shinoda and C.-H. Lee. Structural MAP speaker adaptation using hierarchical priors. In ASRU97, Santa Barbara, USA [ASR97], pages 381–387.

[Sio00] O. Siohan, T. A. Myrvoll, and C.-H. Lee. Joint Maximum A Posteriori Linear Regression for Fast HMM Adaptation. In ASR2000, Paris, France [ASR00], pages 120–123.

[Slo96] T. Sloboda and A. Waibel. Dictionary Learning for Spontaneous Speech Recognition. In ICSLP96, Philadelphia, USA [ICS96], pages 2328–2331.

[Sma] http://www.smartkom.org.

[Spi00] Der Spiegel. Muster im Chaos, June 2000.

[Sto00] A. Stolcke, K. Ries, N. Coccaro, E. Shriberg, R. Bates, D. Jurafsky, P. Taylor, R. Martin, C. Van Ess-Dykema, and M. Meteer. Dialog Act Modeling for Automatic Tagging and Recognition of Conversational Speech. *Computational Linguistics*, 26(3), September 2000.

[Suk96] R. A. Sukkar and C.-H. Lee. Vocabulary Independent Discriminative Utterance Verification for Non-keyword Rejection in Subword based SR. *Transactions on Speech and Audio Processing*, pages 420–429, 1996.

[Suk97] R. A. Sukkar, A. R. Setlur, C.-H. Lee, and J. Jacob. Verifying and Correcting Recognition string Hypotheses using discriminative Utterance Verification. *Speech Communication*, 22:333–342, September 1997.

[The97] E. Thelen, X. Aubert, and P. Beyerlein. Speaker Adaptation in the Philips System for Large Vocabulary Continuous Speech Recognition. In ICASSP97, Munich, Germany [ICA97], pages 1035–1039.

[Tom00] L. Mayfield Tomokiyo. Lexical and Acoustic Modeling of Non-Native Speech in LVCSR. In ICSLP2000, Beijing, China [ICS00].

[Tor95] J. C. Torrecilla, D. Tapias, J. Caminero, and L. Villarrubia. Rejection Techniques Based on Context Independent Subword Units. In Eurospeech95, Madrid, Spain [Eur95], pages 1633–1636.

[Tra99] I. Trancoso, C. Viana, I. Mascarenhas, and C. Teixeira. On Deriving Rules for Nativised Pronunciation in Navigation Queries. In Eurospeech99, Budapest, Hungary [Eur99], pages 195–198.

[Uhr97] C. Uhrik and W. Ward. Confidence Metrics Based On N-Gram Language Model Backoff Behaviors. In Eurospeech97, Rhodes, Greece [Eur97], pages 2771–2774.

[Ven97] Venables and Ripley. *Modern Applied Statistics with S-Plus*. Springer Verlag, 1997.

[Wei95] M. Weintraub. LVCSR Log-Likelihood Ratio Scoring for Keyword Spotting. In ICASSP95, Detroit, USA [ICA95], pages 297–300.

[Wei97] M. Weintraub, F. Beaufays, Z. Rivlin, Y. Konig, and A. Stolcke. Neural - Network Based Measures of Confidence for Word Recognition. In ICASSP97, Munich, Germany [ICA97], pages 887–890.

[Wel98] L. Welling, R. Haeb-Umbach, X. Aubert, and N. Haberland. A Study on Speaker Normalization Using Vocal Tract Normalization and Speaker Adaptive Training. In ICASSP98 [ICA98], pages 797–800.

[Wes96a] M. B. Wesenick. Automatic Generation of German Pronunciation Variants. In ICSLP96, Philadelphia, USA [ICS96], pages 125–128.

[Wes96b] M. B. Wesenick and A. Kipp. Estimating the Quality of Phonetic Transcriptions and Segmentations of Speech Signals. In ICSLP96, Philadelphia, USA [ICS96], pages 129–132.

[Wes98] F. Wessel, K. Macherey, and R. Schlüter. Using Word Probabilities as Confidence Measures. In ICASSP98 [ICA98], pages 225–228.

[Wes00a] M. Wester and E. Fosler-Lussier. A Comparison of Data-Derived Knowledge-Based Modeling of Pronunciation Variation. In ICSLP2000, Beijing, China [ICS00], pages 270–273.

[Wes00b] M. Wester, J. M. Kessens, and H. Strik. Pronunciation Variation in ASR: Which Variation to model. In ICSLP2000, Beijing, China [ICS00], pages 489–492.

[Wil98a] D. Willett, A. Worm, C. Neukirchen, and G. Rigoll. CM for HMM-Based Speech Recognition. In ICSLP98, Sydney, Australia [ICS98], pages 3241–3244.

[Wil98b] G. Williams and S. Renals. Confidence Measures derived from an Acceptor HMM. In ICSLP98, Sydney, Australia [ICS98], pages 831–834.

[Wil98c] G. Williams and S. Renals. Confidence Measures for Evaluating Pronunciation Models. In ROLDUC98 [ROL98], pages 151–155.

[Wil98d] G. Williams and S. Renals. Confidence Measures or Hybrid HMM/ANN Speech Recognition. In ICSLP98, Sydney, Australia [ICS98], pages 3241–3244.

[Wis98] R. Wiseman and S. Downey. Dynamic and Static Improvements to Lexical Baseforms. In ROLDUC98 [ROL98], pages 157–162.

[Wit99a] S. M. Witt and S. J. Young. Off-line Acoustic Modeling of Non-native Accents. In Eurospeech99, Budapest, Hungary [Eur99], pages 1367–1370.

[Wit99b] Silke Maren Witt. *Use of Speech Recognition in Computer-assisted Language Learning*. PhD thesis, University of Cambridge, 1999.

[Wom97] B. D. Womack and J. H. L. Hansen. Stressed Speech Recognition Using Multi-Dimensional Hidden Markov Models. In ASRU97, Santa Barbara, USA [ASR97], pages 404–411.

[Woo99] P. C. Woodland. Speaker Adaptation: Techniques and Challenges. In ASRU99, Colorado, USA [ASR99], pages 85–90.

[Wu98] C.-H. Wu, Y.-J. Chen, and Y.-C. Hung. Telephone Speech Multi-Keyword Spotting Using Fuzzy Search Algorithm and Prosodic Verification. In ICSLP98, Sydney, Australia [ICS98], pages 835–838.

[Yan00a] Q. Yang and J.-P. Martens. Data-driven Lexical Modelling of Pronunciation Variations for ASR. In ICSLP2000, Beijing, China [ICS00], pages 417–420.

[Yan00b] Q. Yang and J.-P. Martens. On the Importance of Exception and Cross-Word Rules for the Data-Driven Creation of Lexica for ASR. In *ProRISC2000 Workshop on Circuits, Systems and Signal Processing*, 2000.

[Zav96] G. Zavaliagkos, R. Schwartz, and J. McDonough. Maximum a Posteriori Adaptation for Large Scale HMM Recognizers. In ICASSP96, Atlanta, USA [ICA96], pages 725–728.

[Zep97] T. Zeppenfeld, M. Finke, K. Ries, M. Westphal, and A. Waibel. Recognition of Conversational Telephone Speech using the Janus Speech Engine. In ICASSP97, Munich, Germany [ICA97], pages 1815–1819.

Index

ISLE, 88
SMARTKOM, 8

minimum classification error training, 59

acoustic models, 10
adaptation
– data, 38
– incremental, 38
– online, 38
– semi-supervised, 73
– speaker, 11
– supervised, 39
– unsupervised, 39
alternate hypothesis, 59
anti-keyword models, 59

Baum-Welch, 28
Bayes theorem, 21

classification error rate, 66
classifier, 60
– Bayesian, 60
– binary decision tree, 60
– LDA, 60
correct classification rate, 66

deletions, 12
discrete cosine transform, 19
DLLR, 46
dynamic programming, 11

Eigenvoices, 41, 106
Expectation-Maximisation, 28

filler models, 59
finite-state grammars, 35
forced alignment, 33, 39
forward procedure, 25
frame, 17

– interval, 18
– length, 17
– shift, 18

Hebbian rule, 64
Hidden Markov Model, 21
– continuous, 24
– discrete, 24
– left–to–right model, 23
– semi-continuous, 25
– training, 27

insertions, 12
IPA, 31

language model, 10, 35
– back-off, 35
– bi-gram, 35
– n-gram, 35
– perplexity, 35
– tri-gram, 35
likelihood ratio, 58, 59

MAP, 41
MAPLR, 45
Mel-frequency cepstral coefficients, 15
minimum verification error training, 59
mixture, 24
ML, 28
MLLR, 42
monophones, 32
multi-layer perceptron, 61
multi-words, 86

neural net
– gradient descent, 65
neural net, 57
– Backpropagation Chunk, 65
– feed forward, 62
– Vanilla Backpropagation, 65
Neyman-Pearson-Lemma, 59

Glossary

Abbreviations and Acronyms

Mathematical Symbols

A Databases and Experimental Settings

A.1 The German Database

The German database was recorded in the sound proof booth at Sony International (Europe) GmbH. The (naive) speakers had to read isolated words that were printed on the screen. The speech was recorded with a high-quality reference microphone, an array microphone and 4 additional, rather simple PC microphones. For the experiments only the speech recorded with the reference microphone was used, unless stated otherwise. The words to be read by the subjects were isolated words and short phrases. They were manually chosen and mainly cover German street and city names and commands. The commands were chosen from various kinds of consumer electronics manuals to cover as many commands as possible for these devices. Also short utterances like 'bitte etwas lauter' (Engl. 'louder please') were recorded. They were treated as one 'word' later during recognition. The address corpus consists of 130 words from which each speaker had to speak 23. The command corpus consists of 527 entries, of which 234 had to be spoken by each speaker. Some more corpora, e.g. digits, pin numbers, spelling, numbers, etc. were also recorded. In total, each speaker had to say approximately 500 utterances. Only the addresses and commands were used for testing. The other corpora were used also for training. The detailed statistics about the German database can be found in Table A.1.

Table A.1. The German speech database (statistics)

	# spks	# utterances (per speaker)	# utterances (total)
Training	185	377	69771
Testing (natives)	15	220	3300
Testing (non-natives)	8	200	1600

The German database also includes a few speakers whose mother tongue was not German, but who were recorded in German nevertheless. These speakers were used in the experiments that were concerned with accented speech. The statistics are also shown in Table A.1 in row three. We used the native speakers to train SI, continuous density, three-state left-to-right HMM models (one for each of the 43 phonemes used), with one mixture density and diagonal covariances.

For the experiments that are described in Chapter 6, 7 and 8 an isolated word recognition task was considered; a grammar in which all words in the vocabulary were equally probable was used. For the experiments using the continuous speech ISLE data base that are described in Chapter 8 a bi-gram language model, that was trained on the data base transcriptions was used.

A.1.1 Pre-processing of the Speech Data

The speech was recorded with the high quality microphone and sampled at 16 kHz with 16 bit resolution. The signal is pre-emphasised and a Hamming window is applied to the signal at an analysis interval of 25 ms. An analysis window is extracted every 10 ms. A Mel-scale filter-bank is used to compute the first twelve MFCC coefficients plus the first and second time derivatives. The energy is not used, but its first and second time derivatives are, making up a 38-component feature vector. In the adaptation experiments only the MFCC components were adapted. This was done to reduce the number of parameters to be adapted. Furthermore, adapting the dynamic coefficients with a small amount of data was found to have an adverse effect, see [Leg95a, Neu95].

A.2 Settings for NN Training and Testing

The training patterns for the NN were generated using the address and command corpora and included OOV words by recognizing approximately 260 utterances (in-vocabulary) per speaker and 260 OOV words, resulting in approximately 520 utterances per speaker. The baseline feature set contained 45 features. We constructed a feature vector for each utterance of the test task and used a NN to classify the recognition result as either correct or wrong. For the NN we used a feed-forward net that consisted of two layers. We used 45 inputs, 10 hidden nodes and 3 output nodes. The number of hidden nodes was varied and we found 10 to be the best choice. The detailed statistics concerning the patterns used for training, evaluation and testing can be found in Table A.2. One pattern corresponds to the features extracted from one utterance.

The data for training the NN were obtained using our standard recogniser with monophone models. We used different data for training the models for the recogniser than we used for training the NN. Since we use some features related to the acoustic score, this was necessary to avoid a possible influence of the training data on the acoustic score if the same models were used. The NN training data was labelled as being correctly recognised or not. So the

Table A.2. Number of patterns used for training, evaluation and testing of the NN classifier

	# patterns	corr	wrong	OOV
train	37718	18859	3022	15837
eval	4250	1795	330	2125
test	16186	5711	2382	8093

target output for the 2 output nodes of the NN were either 0 or 1, '1 0' means the utterance was correctly recognised and '0 1', it was misrecognised. '0 0' and '1 1' were not valid as NN output.

In a second set of experiments we use a NN with 3 output nodes. The first two output nodes have the same meaning as before, the third output node is to indicate, whether a misrecognised word was an OOV word or not ('1' means OOV). So possible outputs (in the training data) are '0 1 1' or '0 1 0' in case of a misrecognition. For correctly recognised words only '1 0 0' is possible. During testing the NN outputs values between 0 and 1. The final decision threshold for the values between 0 and 1 is then 0.5. This means that if the first output is greater than or equal to 0.5 and the second is smaller than 0.5, the utterance is considered as being correctly recognised.

For computing the CER rate of the NN a test set consisting of 35 speakers was used.

A.3 The British English WSJ Database

For the training of the British English monophone models, the British English WSJ database was used. WSJ is a continuous speech (read) database. For the training of the models 44 speakers and 7861 utterances were used.

For the computation of the phoneme bi-gram LM for the phoneme recogniser all phoneme transcription files from 7861 sentences were used. The preprocessing was the same as described in Section A.1. The baseline phoneme recognition rate on the WSJ corpus was 47.3%.

A.4 ISLE Database

The aim of the ISLE project was to develop a language learning tool for second language learners of English. The ISLE database consists of approximately 20 minutes of speech per speaker from 23 German and Italian intermediate learners of English. The sentences were of varying perplexities. In total 11484 utterances were recorded. They were separated into six blocks corresponding to different types of the language learning exercises. Table A.3 gives the number of sentences ('#s') per block together with the linguistic issue that led to the choice of the exercise type and some examples.

Table A.3. ISLE corpus content

Block	#s	ling. issue	ex. type	examples
A B C	27 33 22	Wide vocab. coverage	adaptation	In 1952 a Swiss expedition was sent and two reached a point only three hundred ...
D	81	probl. phones weak phones cons. clusters	mul. choice min. pair	a cup of coffee a mouth - a mouse a railway station
E	63	stress weak forms probl. phones cons. clusters	reading	the convict expressed anger at the sentence the jury took two days to convict him
F	10	weak forms probl. phones	reading selection	I would like chicken with fried potatoes, peas and a glass of water
G	11	weak forms problem phones	reading selection	This year I'd like to visit Rome

B MLLR Results

Table B.1 shows more detailed results for different values of the static weight α. It can be seen that especially in the case of one frame the choice of an inappropriate value of α leads to a dramatic increase in WER. The closer α gets to one, which would correspond to standard MLLR, the bigger is the increase in WER.

Table B.1. WERs using the static weight

α	1 frame		1000 frames	
	addresses	commands	addresses	commands
SI	6.1	13.0	6.1	13.0
0.002	6.1	12.3	6.1	12.8
0.005	6.1	**12.2**	6.1	12.3
0.01	6.1	12.7	6.1	11.9
0.05	7.0	34.6	6.1	10.4
0.1	7.8	49.4	6.1	**10.3**
0.1	7.3	57.2	6.1	**10.3**
0.2	14.5	85.4	6.4	10.4
0.3	26.1	96.6	6.1	10.5
0.4	46.7	99.3	**5.8**	10.6
0.5	68.7	99.4	**5.8**	11.0
0.6	75.7	99.5	6.4	11.5
0.7	84.6	99.6	**5.8**	12.2
0.8	84.9	99.7	**5.8**	12.5
0.9	91.0	99.6	6.1	13.64
relative improvement	-	6.2	4.9	20.8

Some more experiments were conducted to test the performance of the weighted MLLR approach, when the training conditions for the HMM models were different. The results for the α achieving the best results are listed in Table B.2[1]. The results in this table are not directly comparable to each other because of the different training and test sets were used. The different settings for testing were:

- 'non-native speakers': Two Italian speakers, speaking no German, recorded the German address and command corpus. The setting for testing was the one described in 6.2.4.
- 'VM 1 rc': One mixture monophone models were trained using the Verbmobil (VM) database[2] The test set is the one described in 6.2.4. For weighted MLLR one regression class was used.
- 'VM 2 rc': Same as above, using 2 regression classes.
- 'Low cost mic': The test set was recorded with a low cost microphone.
- '4 mix Mono': Four mixture monophone models were trained, the test set was the one described in 6.2.4
- '1 mix tri': one mixture triphones were trained, the test set was the one described in 6.2.4.

Table B.2. WERs for different experiments, using the static weight

experiment	a SI	a MLLR		a impr	k SI	k MLLR		k impr
SI	6.1	-		-	13.0	-		-
non-native spks	20	$\alpha = 0.1$	17.5	12.5	29.2	$\alpha = 0.2$	26.6	8.9
VM 1rc	10.4	$\alpha = 0.5$	7.3	29.8	17.9	$\alpha = 0.2$	11.1	38.0
VM 2rc	10.4	$\alpha = 0.5$	8.7	16.6	17.9	$\alpha = 0.1$	13.0	27.4
low cost mic	6.7	$\alpha = 0.4$	4.6	31.4	11.0	$\alpha = 0.1$	7.1	35.4
4 mix mono	1.2	$\alpha = 0.05$	0.9	25.0	1.6	$\alpha = 0.3$	1.0	37.5
1 mix tri	0.3	$\alpha = 0.1$	0.3	0.0	4.2	$\alpha = 0.3$	2.5	40.5

The results show that the weighted MLLR approach yields improvements for all of the cases. The only exception is the triphone experiment for the address corpus, where the WER could not be reduced further. But please note that the WER is already very low (0.3%) in this case. When the testing conditions differ from the training conditions, usually the SI rates decrease. This is not so much the case when the monophones are used for testing with a low cost microphones (the WER in the SI case increases from 6.1% to 6.7% for the address and for the command corpus the SI results are even slightly

---

[1] for the following values of α the same results as those listed were achieved: for non-native speakers, addresses:$\alpha = 0.2, 0.4$; '4 mix Mono' addresses:$\alpha = 0.09, 0.3, 0.4, 0.5$; commands:$\alpha = 0.4, 0.5, 0.6$; 'one mix tri' addresses:$\alpha = 0.2 - 1.0$
[2] Verbmobil is a German BMBF funded project concerned with spontaneous speech translations for making appointments

better, 11% compared to 13% in the SI). The improvements achieved are a WER reduction of 31.4% and 35.4% for the addresses and commands, respectively. Since only one global regression class was used, the adaptation was rather 'broad'. Therefore this kind of adaptation can also be considered as a channel adaptation. Even if there is only a relatively small mismatch between the training and the testing set, it has an effect on the adaptation, such that using the low cost microphone test data, higher improvements can be achieved than in the initial experiments (shown in Figures 6.3 and 6.4 in Chapter 6). When letting non-Germans speak the test sentences under the same conditions, the mismatch is bigger, since their way of pronouncing the German words deviates a lot from the standard way. This is reflected by the low SI performance of 20% and 29.2% WER for the address and command corpus, respectively. Still, the weighted MLLR approach can improve the results. However, for the commands the improvements are much smaller than for the initial experiments. The reason is that this adaptation is not able to cope with pronunciation errors non-native speakers often make. For instance, they tend to replace sounds of the target language by sounds of their native language. These kinds of errors would need an adaptation that adapts phoneme models specifically. This cannot be achieved with our MLLR approach when one regression class is used. It is however expected that the combination with MAP is helpful in this case, since it achieves a phoneme specific adaptation. However, the amounts of adaptation data were much too small to test this. Also an adaptation of the dictionary as described in Chapter 8 may be necessary to improve the results further.

When the VM data were used for testing, this also means a mismatch between training and testing conditions, since the VM data were recorded externally. The SI performance goes down to 10.4% for the addresses and 17.9% for the commands. MLLR improves the WERs by 29.8% and 38%, respectively. If two regression classes are used, the improvements are smaller, which might again be a result of the reduced amount of adaptation data. This should not play an important role if the adaptation corpus is large enough, but for the small amounts of data considered, it is an important factor.

Also when using more specific models, four mixture monophones or one mixture triphones, for testing, their performance can still be improved by the use of the new MLLR adaptation scheme.

Table B.3 lists the results again using the dynamic weight, one and two regression classes and different numbers of frames. These results correspond to those shown in Figures 6.5 and 6.5 in Chapter 6.

Table B.3. WERs for the command corpus using the dynamic weight

τ	1 regression class					2 regression classes				
	1	1k	2k	5k	10k	1	1k	2k	5k	10k
SI	13.0	13.0	13.0	13.0	13.0	13.0	13.0	13.0	13.0	13.0
1000	18.0	9.9	9.8	10.0	10.0	79.1	10.2	**10.1**	10.9	11.4
1500	15.8	9.9	9.8	10.0	10.0	67.1	10.1	10.3	**10.7**	11.4
2000	13.6	9.9	**9.7**	**9.9**	9.9	55.6	10.2	10.2	**10.7**	11.6
2500	13.2	10.0	9.8	**9.9**	**9.9**	42.9	10.3	10.5	10.8	11.4
3000	13.1	10.1	9.9	10.0	10.0	44.1	10.4	10.4	10.8	11.4
3500	17.0	10.1	9.9	10.1	10.1	34.3	10.5	10.5	10.9	11.4
4000	16.7	**9.9**	9.9	10.1	10.1	24.2	**10.0**	10.6	10.9	11.4
4500	12.6	10.0	9.9	10.1	10.1	36.1	10.6	10.6	11.0	11.3
5000	12.3	10.0	10.0	10.1	10.1	29.2	10.6	10.65	10.9	11.4
5500	12.2	10.1	10.0	10.1	10.1	23.6	10.5	10.7	11.0	**11.3**
6000	12.4	10.1	10.1	10.1	10.1	21.0	10.7	10.8	11.0	11.4
6500	12.8	10.1	10.0	10.2	10.2	22.2	10.8	11.0	11.0	11.4
7000	12.4	10.1	10.0	10.2	10.2	20.2	10.8	11.0	11.0	11.4
7500	12.4	10.1	10.0	10.2	10.2	21.4	10.7	10.9	11.1	11.4
8000	**12.1**	10.1	10.0	10.2	10.2	19.9	10.8	10.9	11.1	11.4
8500	12.3	10.1	9.9	10.2	10.2	18.1	10.9	10.9	11.2	11.4
9000	**12.0**	10.1	10.1	10.2	10.4	17.9	10.9	11.0	11.3	11.5
9500	12.2	10.1	10.1	10.3	10.3	**17.2**	11.0	10.9	11.3	11.5
impr	7.7	23.8	25.4	23.8	23.8	-	23.3	22.5	17.9	13.5

C Phoneme Inventory

We use SAMPA symbols that are proposed for German and British English in [SAM]. In the following, the symbol inventories are given together with examples (orthography and SAMPA transcription). Since some of the English phonemes do not exist in German a mapping to the phonetically closest German phoneme was necessary for some experiments. This mapping is given in the fourth column in Section C.2

C.1 German Symbol Inventory

Symbol	Orthography	SAMPA transcription
Vowels		
I	Ähnlicher	?E:nlIC6
i	binär	binE:6
i:	Dienstag	di:nsta:g
U	Bewertung	b@ve:6tUN
u	Musik	muzi:k
u:	gutem	gu:t@m
Y	Informationssystem	?InfO6matsjo:nszYste:m
y	Büro	byRo:
y:	Vorführung	fo:6fy:RUN
E	welches	vElC@s
e	Reflex	ReflEks
e:	Bewertung	b@ve:6tUN
E:	Ähnlich	?E:nlIC
O	Donnerstag	dOn6sta:g
o	Poet	poe:t
o:	Montag	mo:nta:g
9	zwölfter	tsv9lft6
2	möbliert	m2bli:6t
2:	Regisseur	ReZIs2:R
a	achter	?axt6
a:	Dienstag	di:nsta:g
@	Ähnliches	?E:nlIC@
6	Ähnlicher	?E:nlIC6
Diphthongs		
aI	Preise	pRaIz@
aU	laufe	laUf@
Y	anläuft	?anlOYft
Semivowel:		
j	jeglichem	je:klIC@m

Symbol	Orthography	SAMPA transcription
Consonants		
Plosives		
p	Platz	plats
b	Bewertung	b@ve:6tUN
t	Bewertung	b@ve:6tUN
d	Dienstag	di:nsta:g
k	Kasse	kas@
g	Donnerstag	dOn6sta:g
Frikatives		
f	Film	fIlm
v	Wege	ve:g@
z	Preise	pRaIz@
S	Spielfilm	Spi:lfIlm
Z	Regisseur	Re:ZIs2:R
C	Ähnliche	?E:nlIC@
x	achtem	?axt@m
h	Hauptdarsteller	haUptda:6StEl6
Liquides		
l	Ähnliche	?E:nlIC@
R	Freitage	fRaIta:g@
Nasals		
n	Ähnlicher	?E:nlIC6
m	Filmes	fIlm@s
N	Längen	lEN@n
Glottal stop		
?	Ähnliche	?E:nlIC@

C.2 English Symbol Inventory

Symbol	Orthography	SAMPA	German
Consonants and semi vowels			
T	thin	TIn	s
D	this	DIs	z
w	will	wIl	v
Consonant clusters word initial			
sm	smile	smaIl	sm
sn	snake	sneIk	sn
sw	sweater	swet@	sv
sp	sprite	spraIt	sp
st	star	stA:	st
English lenis word final			
b	cab	c{b	p
d	bad	b{d	t
g	dog	dO:g	k
z	is	Iz	s
Vowels			
{	pat	p{t	E
Q	pot	pQt	O
V	cut	cVt	a
3:	furs	f3:z	96
A:	stars	stA:z	a:
O:	cause	kO:z	o:
Diphthongs			
eI	mail	meIl	e:
OI	noise	nOIz	OY
@U	nose	n@Uz	o:
I@	fears	fI@z	i:6
e@	stairs	ste@z	E:6
U@	cures	kjU@z	u:6

C.3 Manually Derived Pronunciation Rules for the ISLE Corpus

In this section the manually derived pronunciation rules for German and Italian second learners of English of the ISLE data base are listed. These were used for the experiments in Section 8.3. A more detailed description of the related experiments can be found in [Sah01]. How often the phoneme obeying the rule was involved in errors is reported as 'occurrence'.

Pronunciation Rules for German Speakers

- Word final devoicing: German speakers have difficulties in pronouncing final voiced obstruents in English (in German all final obstruents are voiceless).
  e.g., cause /k o: z/ → /k o: s/
  Occurrence: 14%
- Vowel Reduction: This takes place in English in unstressed syllables. The vowel that is mostly affected is /@/. In German unstressed vowels are, with few exceptions, pronounced with their full quality.
  e.g., another /@ n V D 3:/ → /e n V D 3:/
  Occurrence: 21.7%
- Vowel /{/: German speakers replace /{/ by the nearest vowel /E/ in words like "shall", "pan". These mispronunciations are typical examples of intralingual mistakes.
  e.g., planing /p l { n I N/ → /p l e n I N/
  Occurrence: 3.6%
- /w/: German speakers have difficulties in pronouncing /w/, as this sound does not exists in modern German. They often replace /w/ by /v/ in words like "workforce". Some German speakers also replace /v/ by /w/. This is termed hypercorrection,
  e.g., very well /v { r i: v E l/ → /w { r i: w E l/
  Occurrence: /w/ to /v/: 2.9%, /v/ to /w/: 1.0%
- /Ng/-problem: This consonant combination does not exist in German. Therefore the second consonant /g/ is very often deleted at morpheme boundaries,
  e.g., finger /f I N g 3:/ → /f I N 3:/
  Occurrence: 0.2%
- /D/-problem: German (also Italian) speakers have great articulatory difficulty in pronouncing the dental fricative /D/, especially in combination with other fricatives. They replace it by the plosive consonant /d/
  e.g., other /V D 3:/ → /V d 3:/.
  Occurrence: 0.8% for German, 5% for Italian speakers

Pronunciation Rules for Italian Speakers

- Vowel Lengthening: Since Italian speakers have difficulties in perceiving the distinction between the two sounds, often /I/ is replaced by /i:/.
  e.g., river /R I v @/→ /R i: v @/
  Occurrence: 10%
- /@/-problem: Most Italian speakers append a /@/ at word final consonants.
  e.g., able /eI b @ l/ → /eI b @ l @/
  Occurrence: 13.5%

– /h/-problem: Italian speakers have difficulties in pronouncing /h/ in word initial position and thus it is often deleted.
e.g., height /h eI t/ → /eI t/
Occurrence: 1.7%

The vowel reduction turned out to be a very important rule, however its application to the baseline dictionary was not as straightforward as for the other rules. While for the other rules a mapping on the phoneme level was directly possible, adding the variants that were due to the vowel reduction required a back off to the orthography. In English many vowels when being reduced are reduced to /@/ in unstressed syllables. As stated above, especially German but also Italian speakers have difficulties in realizing this /@/ and tend to pronounce it in its full quality. The actual realisation depends on the grapheme and its context that was found in the orthography.

Lecture Notes in Artificial Intelligence (LNAI)

Lecture Notes in Computer Science